Practical Inverse Analysis in Engineering

CRC MECHANICAL ENGINEERING SERIES

Edited by Frank A. Kulacki, University of Minnesota

Published

Entropy Generation Minimization
Adrian Bejan, Duke University

The Finite Element Method Using MATLAB
Young W. Kwon, Naval Postgraduate School
Hyochoong Bang, Korea Aerospace Research Institute

To be Published

Fundamentals of Environmental Discharge Modeling
Lorin R. Davis, Oregon State University

Mechanics of Composite Materials
Autar K. Kaw, University of South Florida

Nonlinear Analysis of Structures
M. Sathyamoorthy, Clarkson University

Mechanics of Solids and Shells
Gerald Wempner, Georgia Institute of Technology

Viscoelastic Solids
Roderic Lakes, University of Iowa

Practical Inverse Analysis in Engineering

David M. Trujillo
Henry R. Busby

CRC Press
Boca Raton New York

Acquiring Editor: Bob Stern
Project Editor: Gail Renard
Cover Design: Dawn Boyd
PrePress: Kevin Luong

LIMITED WARRANTY

CRC Press LLC warrants the physical diskette(s) enclosed herein to be free of defects in materials and workmanship for a period of thirty days from the date of purchase. If within the warranty period CRC Press LLC receives written notification of defects in materials or workmanship, and such notification is determined by CRC Press LLC to be correct, CRC Press LLC will replace the defective diskette(s).

The entire and exclusive liability and remedy for breach of this Limited Warranty shall be limited to replacement of defective diskette(s) and shall not include or extend to any claim for or right to cover any other damages, including but not limited to, loss of profit, data, or use of the software, or special, incidental, or consequential damages or other similar claims, even if CRC Press LLC has been specifically advised of the possibility of such damages. In no event will the liability of CRC Press LLC for any damages to you or any other person ever exceed the lower suggested list price or actual price paid for the software, regardless of any form of the claim.

CRC Press LLC specifically disclaims all other warranties, express or implied, including but not limited to, any implied warranty of merchantability or fitness for a particular purpose. Specifically, CRC Press LLC makes no representation or warranty that the software is fit for any particular purpose and any implied warranty of merchantability is limited to the thirty-day duration of the Limited Warranty covering the physical diskette(s) only (and not the software) and is otherwise expressly and specifically disclaimed.

Since some states do not allow the exclusion of incidental or consequential damages, or the limitation on how long an implied warranty lasts, some of the above may not apply to you.

DISCLAIMER OF WARRANTY AND LIMITS OF LIABILITY: The author(s) of this book have used their best efforts in preparing this material. These efforts include the development, research, and testing of the theories and programs to determine their effectiveness. Neither the author(s) nor the publisher make warranties of any kind, express or implied, with regard to these programs or the documentation contained in this book, including without limitation warranties of merchantability or fitness for a particular purpose. No liability is accepted in any event for any damages, including incidental or consequential damages, lost profits, costs of lost data or program material, or otherwise in connection with or arising out of the furnishing, performance, or use of the programs in this book.

Library of Congress Cataloging-in-Publication Data

Trujillo, David M.
Practical inverse analysis in engineering/David M. Trujillo, Henry R. Busby.
p. cm. — (CRC mechanical engineering series)
Includes bibliographical references and index.
ISBN 0-8493-9659-X (alk. paper)
1. Engineering—Mathematical models—Data processing. 2. Dynamic programming.
I. Busby, Henry R. II. Title. III. Series: Advanced topics in mechanical engineering series.
TA342.T78 1997
620′. 001′519703—dc21 97–16868
CIP

International Standard Book Number 0-8493-9659-X
Library of Congress Card Number 97-16868
Printed in the United States of America 1 2 3 4 5 6 7 8 9 0
Printed on acid-free paper

To our wives—Lynne and Nancy

About the Authors

David M. Trujillo received his B.S. and M.S. degrees in engineering from Arizona State University in 1961 and 1963, respectively. Dr. Trujillo worked at Rocketdyne, Canoga Park, California, for several years before entering the University of Southern California, where he received his Ph.D. in mechanical engineering in 1971. He has experience in a variety of disciplines, including shock and vibration, inelastic stress analysis, and heat transfer.

He also worked at TRW, Redondo Beach, and Holmes & Narver, Orange, California, before becoming a consultant in 1980 where he developed several finite element codes for ground water flow and inverse heat conduction. He joined Hughes Electronics in 1990 where he is currently a senior scientist supporting spacecraft analysis.

A member of ASME, Dr. Trujillo has authored or coauthored more than 25 papers in the areas of numerical methods and inverse problems.

Henry R. Busby earned an undergraduate degree in mechanical engineering at California State University, Long Beach, and his M.S. and Ph.D. at the University of Southern California. He is currently Professor of Mechanical Engineering at The Ohio State University where he has taught machine design, computer-aided design, and composites at the undergraduate and graduate levels since 1982. He is coauthor of *Introductory Engineering Modeling Emphasizing Differential Models and Computer Simulations.*

In addition to his career in academia, Professor Busby has compiled more than 15 years of experience in industry, as well as acting as a consultant with government and various engineering firms.

He has publishedmore than 30 papers in technical journals and is a member of the American Society of Mechanical Engineers, American Academy of Mechanics, Society of Experimental Mechanics, and the Society of Industrial and Applied Mathemematics.

PREFACE

The current revolution in computer technology is producing larger, faster, and inexpensive computers with almost no limits in sight. Today's personal computers can even be easily modified to accept measured data directly from transducers. This technology has made it possible for engineers and scientists to construct more realistic mathematical models of physical processes. This book addresses an important area of engineering that will become even more significant in the future — the combining of measurements with engineering models.

It is the purpose of this book to present pragmatic mathematical tools that can be used to bridge the disciplines of real-world measurements and mathematical models. Traditionally these disciplines are rarely discussed together, and understandably so. Almost all engineering analysis is based on models that have nice smooth mathematical properties, often possessing derivatives of very high order. On the other hand, any measurement is viewed as having some noise, and the question of derivatives is not even contemplated. Even textbooks on numerical analysis rarely devote more than one or two pages to estimating derivatives of real-world measurements. The problem is dismissed as *ill-conditioned*. Ill-conditioned refers to a situation in which the solution is extremely sensitive to the data. However, combining measurements with engineering models almost always results in an ill-conditioned problem.

The inverse problem is one type of ill-conditioned or ill-posed problem. In a typical direct problem, one is given a model, the initial conditions of the state variables, the forcing terms, and is asked to produce a solution. In the general inverse problem, one is given a model and measurements of some state variables and is asked to estimate the initial conditions, the forcing terms, and the rest of the state variables.

This book will show how to treat these ill-conditioned problems with the powerful and general theory of dynamic programming. In addition, another theory — generalized cross validation — will be discussed and shown to be a useful partner in handling real data. The combination offers objective and pragmatic tools for obtaining solutions to these inverse problems. This book is intended for senior undergraduates and graduate engineering students and for the practicing engineer.

Chapter 1 presents the elements of discrete dynamic programming, starting with the simplest examples. This is so the reader may gain a full understanding of the method. Next, a scalar dynamic model is discussed as a precursor to the ultimate goal — large-scale engineering models. These will require vector-matrix notation. Chapter 2 discusses large- scale dynamic models and the various methods for representing these with discrete models. The general inverse problem is discussed in Chapter 3, which also shows how generalized cross validation can be used to obtain useful solutions to ill-conditioned inverse problems.

Chapters 4, 5, and 6 contain specific applications of the methods to engineering problems in heat transfer, structural dynamics, and the smoothing and differentiation of noisy data. Realistic models are also nonlinear, so in Chapter 7 several methods are presented which can be used to solve these systems. Finally, Chapter 8 shows how systems can be constructed from measurements.

This book grew out of numerous technical papers by the authors over the past 20 years. Hopefully, the experience gained in transforming mathematical ideas into practical computer programs is reflected in the book. It is also hoped that this book will allow engineers and scientists to improve their models, to obtain useful information from tests, and to design better experiments and instrumentation. We have taken the first step in this area, mostly from the mathematical side, with the purpose that the next steps will be easier.

We are indebted to our colleagues Clark Dohrmann and Donald Tandy, who contributed research in the areas of smoothing and differentiation. We also wish to express our thanks to Gail Renard of CRC Press for her editorial assistance and preparation of the final manuscript.

David M. Trujillo
Henry R. Busby
March 1997

Contents

Chapter 1

DYNAMIC PROGRAMMING

1.1 INTRODUCTION

The mathematical theory of dynamic programming is unlike any other encountered in engineering mathematics. While it covers a wide range of optimization subjects, we will focus only on a small yet important part. This includes systems that are governed by ordinary and partial differential equations, the basis of most engineering models. In order to fully explain dynamic programming we will discuss the simplest optimization problem, the minimum path between two points. This example will illustrate the salient features of the theory, which will then be formalized and applied to a first-order dynamic system. The extension to multidimensional systems will then be made. Once the idea is understood it will seem fairly simple, yet since it is unlike any other it is sometimes difficult to grasp. The approach we have taken is to build on concepts by starting very simply and then proceeding to the next more general case.

1.2 THE SIMPLEST EXAMPLE

This example is used in many texts (Bellman and Dreyfus, 1962) and justifiably, because it illustrates all of the ingredients necessary for applying dynamic programming. Figure 1.1 shows a grid representing the costs of traveling from one point on the grid to another. The optimization problem is to minimize the total cost of traveling from point A to point B. The choice at any point will be limited to either going to the right or up.

An obvious solution is to try every path between A and B and select the minimum. This is an easy and trivial task for a

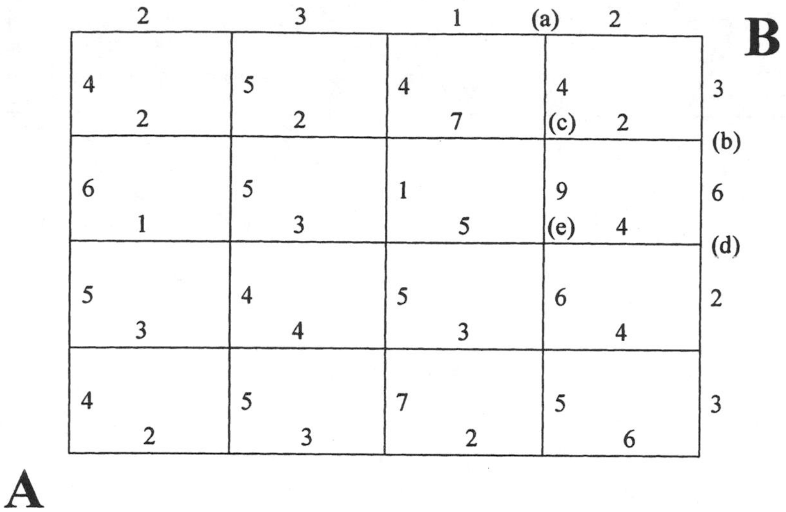

FIGURE 1.1
Cost layout for simple example.

computer. The only drawback is that as the size of the grid becomes larger this approach quickly becomes a formidable task even for the fastest computer. A better solution exists and has some additional benefits which will be pointed out.

The first step is to start at the end point B and solve the problem backward, ending at point A. This is a trait of dynamic programming that will occur in all of the applications. For point (a) in Figure 1.1 the minimum cost to travel to B is 2 and there is no choice to be made; one must travel to the right. Thus the minimum cost to go from point (a) to B can be expressed as

$$F(a) = 2$$

and the optimal policy as

$$G(a) = \text{go to the right}$$

This is another distinct feature of dynamic programming; there are *two* functions that comprise the solution, an optimal cost function *and* an optimal policy function. The functions for point (b) are similarly found to be

$$F(b) = 3 \text{ and } G(b) = \text{go up}$$

Now consider point (c) which offers two choices in the policy, go to the right or go up. The optimum choice is the one that minimizes the following

$$F(c) = \text{minimize}\begin{cases} 2 + F(b) \\ 4 + F(a) \end{cases}$$

The top expression is applicable when going to the right, and the bottom one by going up. In this case the optimal policy is to go to the right. This gives the optimal cost and policy for point c as

$$F(c) = 5 \text{ and } G(c) = \text{go to the right}$$

The most important feature of the minimization is that it involves a choice between two previously determined *optimal* functions F(a) and F(b). This is the multistage feature of dynamic programming. It has successfully divided the overall problem into a succession of simpler ones. The minimization is performed over the sum of a local cost and the remaining optimal cost resulting from changing the grid point due to the local decision.

The optimum cost for point (d) is simple since there is no choice to be made

$$F(d) = 6 + F(b) \text{ and } G(d) = \text{go up}$$

Point (e) is determined from

$$F(e) = \text{minimize}\begin{cases} 2 + F(d) \\ 4 + F(c) \end{cases}$$

This gives

$$F(e) = 13 \text{ and } G(e) = \text{go to the right}$$

The rest of the points can easily be calculated. The reader is strongly encouraged to complete the rest of the calculations. Continuing until the last point (A) is reached results in the following

$$F(A) = \text{minimize}\begin{cases} 4 + F(y) : F(y) = 17 \\ 2 + F(x) : F(x) = 20 \end{cases}$$

The minimization gives the optimal cost and the optimal policy for point A as

$$F(A) = 21 \text{ and } G(A) = \text{go up}$$

This completes the backward phase of the solution. The forward phase starts at point A and, since G(A) = go up, this takes us to point (y) where G(y) = go up, etc. The forward phase simply follows the optimal policies determined during the backward phase. Figure 1.2 shows the optimal path from A to B.

The dynamic programming solution has some additional benefits over the brute force approach of calculating every possible path between A and B. The first is that suppose one did

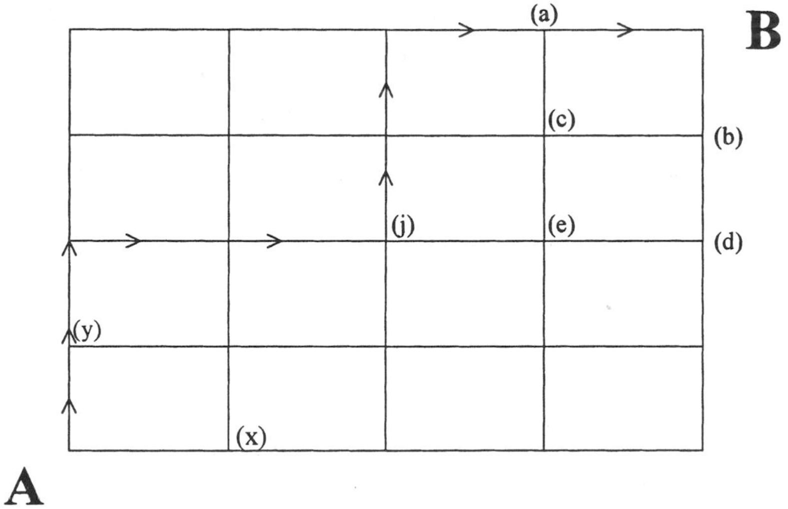

FIGURE 1.2
The optimal path.

not want to start at point A but at point (x). The optimal cost and policy for this case have already been determined. F(x) is the optimal cost, and G(x) is the optimal policy starting from point (x). In fact, one has calculated the optimal cost and policy for every point in the grid. This is in direct contrast to the brute force approach, which finds only the one solution from A to B.

The second benefit is one of stability, which is closely related to the fact that the optimal cost has been found for every point. The optimum path from A to B found by dynamic programming will be the same as the one found by the brute force approach. However, consider the case where the path is perturbed while being traveled. That is, suppose for some reason that at point (j) it becomes impossible to go up and one must go to the right. This will put one at point (e). Clearly, the minimum cost can no longer be achieved. However, the dynamic programming solution provides the next best thing, the optimal cost and policy starting at point (e). This general feature is implicit in the dynamic programming solution. The single optimal path being sought is embedded in many suboptimal solutions, and thus perturbations are easily accepted and cause no difficulties. This is the essence of stability. This feature will become more important when large dynamic systems are being modeled.

1.3 BELLMAN'S PRINCIPLE OF OPTIMALITY

In the previous example, the functional equation being used was actually an application of the Principle of Optimality developed by mathematician Richard Bellman.

> **Principle of Optimality**: *An optimal policy has the property that whatever the initial state and initial decision are, the remaining decisions must constitute an optimal policy with regard to the state resulting from the first decision.*

The mathematical application of this principle will serve as the basis of most of the subsequent chapters in this book. As with most powerful ideas, after it is stated it immediately becomes intuitive. In recalling the simple example previously discussed, once the solution is presented it is difficult to accept any other.

The next sections will apply the principle of optimality to a scalar dynamic system and then extend the concepts to the general multidimensional system. Dynamic programming has been applied to a much wider variety of problems than those discussed here. The reader is encouraged to review the bibliography for more information, especially Bellman (1957), Bellman and Dreyfus (1962), and Larson and Casti (1982).

1.4 FIRST-ORDER DYNAMIC SYSTEM

The general discrete first-order system is governed by

$$x_{j+1} = mx_j + g_j \tag{1.4-1}$$

Suppose some measurements have been taken on x; let these be denoted by d_j, j = 1,N. The optimization problem is to determine the forcing terms g_j which will be applied to the system and the initial condition on x such that the following function is minimized:

$$E = \sum_{j=1}^{N} \left(d_j - x_j\right)^2 \tag{1.4-2}$$

This is a common problem in the estimation of state variables. If m = 1 this becomes a problem of estimating the displacement x_j and the velocity g_j for a constant timestep h, which is set to 1 for this example. Since measurements are involved, the data will always contain errors or noise, so it is fruitless to exactly match the data. Least squares is one of the more useful methods for matching data. While there are many underlying assumptions involved in justifying its use, most practical engineering problems use this criterion for a best-match solution. However, even this least squares criterion is not sufficient, because a mathematical solution that will minimize the above function E will end up with the model exactly matching the data — a situation that is to be avoided. This is where the regularization method enters. By adding a term to the above least squares error function

$$E = \sum_{j=1}^{N} \left(d_j - x_j\right)^2 + bg_j^2 \qquad (1.4\text{-}3)$$

one can control the amount of smoothness that occurs in the solution by varying the parameter b. This method is sometimes referred to as Tikhonov's method. What is now required of the solution is to best match the data (the first term of Equation 1.4-3), but also to have some degree of smoothness (the second term). This immediately brings up the question of what should be the value of the smoothing parameter b. Fortunately, there exists a method to determine the optimum value of b. This method is called *generalized cross validation* and will be discussed in a later chapter. It is sufficient at this point to notice that if b is zero, or very small, the data will be followed exactly, and if b is very large, the system will remain stationary. Thus it follows that there exists a value of b between these two extremes that will produce the best estimate.

The dynamic programming solution to the above problem begins by defining the following function:

$$f(c) = \min_{g_j} E_n(c, g) \qquad (1.4\text{-}4)$$

This function represents the minimum value of E starting at any stage $j = n$ with $x_n = c$ and simulating the system to the end ($j = N$) with the *optimal* g_j's. The important items to emphasize are that c is considered to be arbitrary and that n can represent any value between 1 and N. Applying the principle of optimality leads to the following recurrence formula:

$$f_{n-1}(c) = \min_{g_{n-1}} \left[\left(d_{n-1} - c\right)^2 + bg_{n-1}^2 + f_n\left(mc + g_{n-1}\right) \right] \qquad (1.4\text{-}5)$$

This situation is similar to that encountered in the simple example given in Section 1.2, in that the minimum at any point is determined by selecting the decision (g_{n-1}) to minimize the immediate cost (first and second terms of Equation 1.4-5) and the remaining cost resulting from the decision (the third term).

The decision will result in the next state mc + g_{n-1}. Notice that the minimum is performed over a previously determined optimal function f_n, i.e., going backward.

It is important to notice that the recurrence formula represents two functions, f_n and the optimal forcing terms g_n. The solution to the above formula is obtained by starting at the end of the process n = N and working backward to n = 1. At the end point, n = N, the minimum value of E is given by

$$f_N(c) = \min_{g_N}\left[(d_N - c)^2 + bg_N^2\right] \tag{1.4-6}$$

The forcing term g_N that minimizes the above expression is simply $g_N = 0$, which gives that

$$f_N(c) = (d_N - c)^2 \tag{1.4-7}$$

This is interpreted to mean that if there were only one stage and $x_N = c$, then the error is related only to d_N. This is the advantage of starting at the end point; the optimum solution is easy to determine at this point.

Now consider that the initial state c was started at n = N – 1, one step back. The recurrence formula now becomes

$$f_{N-1}(c) = \min_{g_{N-1}}\left[(d_{N-1} - c)^2 + bg_{N-1}^2 + f_N(mc + g_{N-1})\right] \tag{1.4-8}$$

Since f_N has previously been determined by Equation 1.4-7 for an arbitrary argument, this equation becomes

$$f_{N-1}(c) = \min_{g_{N-1}}\left[(d_{N-1} - c)^2 + bg_{N-1}^2 + (d_N - mc - g_{N-1})^2\right] \tag{1.4-9}$$

Performing the indicated minimization gives the *optimum* value of the forcing term (denoted by an asterisk)

$$g_{N-1}^* = (d_N - mc)/(b+1) \tag{1.4-10}$$

Substituting this expression for g_{N-1} in Equation 1.4-9 gives

$$f_{N-1}(c) = (d_{N-1} - c)^2 + b(d_N - mc)^2/(b+1)^2 + \left[d_N - mc - (d_N - mc)/(b+1)\right]^2 \tag{1.4-11}$$

At this point we actually have a complete solution for one step. If we started at N – 1 with $x_{N-1} = c$, then the optimal choice for g_{N-1} is calculated with Equation 1.4-10. This would take us to x_N given by

$$x_N = mc + g^*_{N-1} \tag{1.4-12}$$

The above process is continued, working backward, and at each step computing f_n and g^*_n. A direct algebraic approach would soon generate unwieldy expressions. Fortunately there is a structure in the process that simplifies the calculations tremendously. This structure is to notice that every expression for $f_n(c)$ is quadratic in c. That is, it can be proven inductively that for any n

$$f_n(c) = r_n c^2 + s_n c + q_n \tag{1.4-13}$$

This can be seen to be true for Equations 1.4-7 and 1.4-11, n =N and n = N – 1. Equation 1.4-13 can now be used to derive recursive relations between r_n, s_n, q_n and r_{n-1}, s_{n-1}, q_{n-1}. Returning to Equation 1.4-5 and using the quadratic expression for f_n gives

$$r_{n-1}c^2 + s_{n-1}c + q_{n-1} = \min_{g_{n-1}}\left[(d_{n-1} - c)^2 + bg^2_{n-1} + r_n(mc + g_{n-1})^2 + s_n(mc + g_{n-1}) + q_n\right] \tag{1.4-14}$$

Performing the indicated minimization gives

$$2bg^*_{n-1} + 2r_n(mc + g_{n-1}) + s_n = 0$$

or

$$g^*_{n-1} = (-2r_n mc - s_n)/(2b + 2r_n) \tag{1.4-15}$$

The optimal forcing term is linear in c. This expression for the optimal forcing term can now be substituted back into Equation 1.4-14 and then if the coefficients of like terms of c are equated, the following recurrence formulas can be derived:

$$r_{n-1} = 1 + bm^2 r_n/(b + r_n) \tag{1.4-16}$$

$$s_{n-1} = -2d_{n-1} + mbs_n/(b + r_n) \tag{1.4-17}$$

$$q_{n-1} = q_n + d_{n-1}^2 - s_n^2/4(b + r_n) \tag{1.4-18}$$

These equations are solved backward starting with the initial conditions given by Equation 1.4-7, which are

$$r_N = 1; \quad s_N = -2d_N; \quad q_N = d_N^2 \tag{1.4-19}$$

The forward solution starts at n = 1 where the minimum of E is given for any arbitrary value of c, mainly

$$f_1(c) = r_1 c^2 + s_1 c + q_1 \tag{1.4-20}$$

At this point we must choose the optimal initial condition, which is given by minimizing f_1 with respect to c. This gives the optimal initial condition

$$c_1^* = -s_1/(2r_1) \tag{1.4-21}$$

The forward solution is now calculated using Equation 1.4-15 to calculate the optimum forcing term g_1^* and Equation 1.4-1 to advance the state of the system. The only storage requirements are for the coefficients r_n and s_n.

For a numerical example, let us solve a problem with 11 steps. The parameter m will be set to 1 and the smoothing parameter will be chosen as b = 0.2. The data were generated by sampling the function 1 – cos(πt) from zero to 1 with a sampling increment of 0.10. Some *noise* was added to the data to simulate a real measurement. The derivative of the above function is πsin(πt).

TABLE 1.1

Results of Backward Calculations (b = 0.2)

STEP	d	r	s	q
1	–0.010	1.171	–0.018	0.100
2	0.090	1.171	–0.257	0.112
3	0.180	1.171	–0.528	0.155
4	0.440	1.171	–1.154	0.365
5	0.760	1.171	–1.876	0.813
6	1.000	1.171	–2.437	1.319
7	1.220	1.171	–2.998	1.958
8	1.600	1.171	–3.825	3.138
9	1.800	1.171	–4.284	3.926
10	2.010	1.167	–4.677	4.687
11	1.970	1.000	–3.940	3.881

The results of the backward calculations are shown in Table 1.1 for a smoothing parameter b equal to 0.2. The initial conditions at step 11 were calculated using Equation 1.4-19. One very important observation on the behavior of r_n is that it approaches a steady-state value. This fact will be very important for the multidimensional case in which r_n will represent a matrix and whose calculation will dominate the computational effort. By monitoring this convergence, a significant savings can be achieved. Equation 1.4-16 is known as the *discrete Riccati equation*.

The forward solution begins by calculating the optimum initial condition using Equation 1.4-21:

$$c_1^* = -s_1/2r_1 = 0.0077$$

which has been rounded off to 0.008 in Table 1.2. With this value, the optimum value of the forcing term can be calculated using Equation 1.4-15:

$$g_1^* = (-2r_2mc - s_2)/(2b + 2r_2) = 0.087$$

The next value of x can now be calculated using Equation 1.4-1:

$$x_2 = x_1 + g_1 = 0.0077 + 0.087 = 0.0947$$

TABLE 1.2

Results of the Forward Calculations (b = 0.2)

STEP	x	d	g
1	0.008	–0.010	0.087
2	0.095	0.090	0.112
3	0.207	0.180	0.244
4	0.451	0.440	0.299
5	0.750	0.076	0.249
6	0.998	1.000	0.241
7	1.239	1.220	0.337
8	1.576	1.600	0.217
9	1.793	1.800	0.181
10	1.973	2.010	–0.003
11	1.971	1.970	0.000

The rest of the forward calculations are shown in Table 1.2, where the data have been included for ease of comparison.

While the effect of the smoothing parameter is relatively small on the estimation of x, it greatly affects the estimation of g. This can be seen in Figure 1.3, which shows the optimal

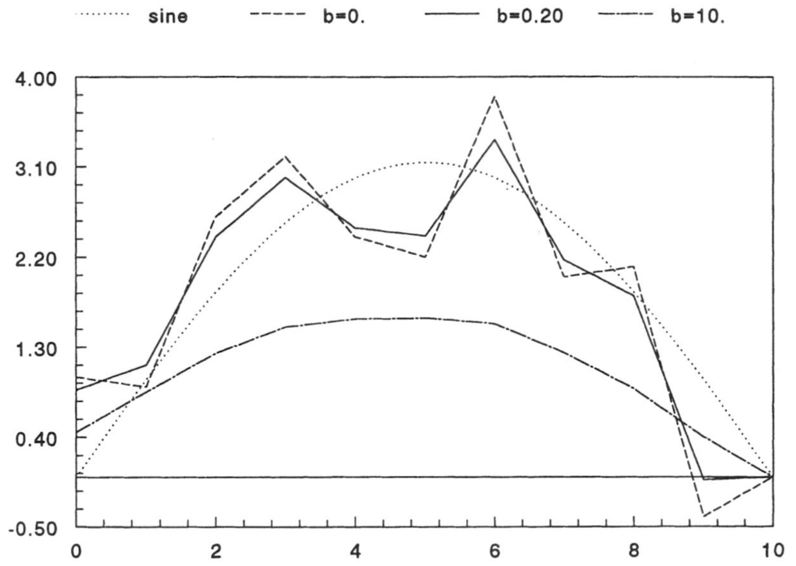

FIGURE 1.3
Comparison of g (modified for stepsize) for various values of b.

estimation of g for various values of the smoothing parameter. It is clear that as b is increased, the values of g become smaller, and for very large values, g would become almost zero and x would become a constant equal to the average value of the data.

It is interesting that the variables q_n do not enter into the estimation of either x or g in the forward calculations. Also, they are not needed to calculate the value of the minimum error since this can be done during the forward calculations using the optimal values of x and g as they are being calculated. Indeed, one useful calculation to perform during the forward calculations is the mean square error between the data and the estimation. In the multidimensional case, the calculation of the q_n's will usually be disregarded.

Although this example has only 11 points it is clear that the above equations could be programmed to easily solve a 1000-step problem. Only the sequences r_n and s_n have to be stored since these are required by Equation 1.4-15. When one realizes that this is equivalent to minimizing a function with 1000 variables, it becomes clear that dynamic programming is the only method to use for these types of problems. In Chapter 6, more practical models will be introduced to smooth noisy data and to estimate their first and second derivatives.

1.5 GENERAL MULTIDIMENSIONAL SYSTEM

The general multidimensional system that will serve as the basic model for the rest of the book is expressed in discrete form as

$$\mathbf{x}_{j+1} = \mathbf{M}\mathbf{x}_j + \mathbf{P}\mathbf{g}_j \tag{1.5-1}$$

where j = 1,N and

- **x** is an (m × 1) vector representing the state variables
- **g** is an (m_g × 1) vector representing the forcing variables
- **M** is an (m × m) system matrix
- **P** is an (m × m_g) matrix relating the forcing terms to the state variables

There are many methods that can be used to represent a continuous system with the matrices **M** and **P**. These will be discussed in Chapter 2. The initial condition is given by the vector **c** so that

$$\mathbf{x}_1 = \mathbf{c} \tag{1.5-2}$$

The problems we are interested in solving involve those in which some measurements have been taken on some of the state variables **x**. The most common case is where the number of measurements m_z is much less than the number of state variables m. Let these measurements be denoted with the vector **d**, where **d** is an $(m_z \times 1)$ vector representing the measurements. A more general relationship that relates the state variables to the measurements $\mathbf{d}_j$ can be expressed as

$$\mathbf{z}_j = \mathbf{Q}\mathbf{x}_j \tag{1.5-3}$$

where

Q = an $(m_z \times m)$ matrix
$\mathbf{z}_j$ = an $(m_z \times 1)$ vector

Now $\mathbf{z}_j$ can be compared directly to $\mathbf{d}_j$. This more general form allows one to relate the measurement to a linear combination of the state variables. This can arise in the analysis of spring-mass dynamic systems where the measurements are usually accelerations, while the state variables are displacements and velocities.

The problem is to find the unknown forcing terms $\mathbf{g}_j$ that will cause the system (Equation 1.5-1) to best match the measurements $\mathbf{d}_j$. The mathematical representation of a best match is to minimize the least squares error between $\mathbf{d}_j$ and $\mathbf{z}_j$. This is expressed in matrix-vector notation with the inner product of two vectors (.,.). Chapter 2 contains a brief description of matrix-vector notation and definitions.

The general least squares error is now expressed as

$$E(\mathbf{c},\mathbf{g}_j) = \sum_{j-1}^{N} \left(\mathbf{z}_j - \mathbf{d}_j, \mathbf{A}(\mathbf{z}_j - \mathbf{d}_j)\right) - + \left(\mathbf{g}_j, \mathbf{B}\mathbf{g}_j\right) \tag{1.5-4}$$

where $\mathbf{A}(m_z \times m_z)$ and $\mathbf{B}(m_g \times m_g)$ are symmetric positive definite matrices that provide the flexibility of weighting the measurements and the forcing terms. The exact specification of these matrices will require some additional information that is problem dependent. Usually **A** is the identity matrix and **B** is a diagonal matrix. As discussed in Section 1.3, the addition of the term $(\mathbf{g}_j,\mathbf{B}\mathbf{g}_j)$ is crucial to obtaining smooth and reasonable results. It is known as the *regularization term* and is sometimes referred to as *Tikhonov's method*. Another way to interpret Equation 1.5-4 is that we seek a solution that best matches the data (the first term of the equation), but also has some degree of smoothness (the second term). This immediately brings up the question as to the value of the smoothing matrix **B**. Fortunately, there exists a method that can be used to estimate the optimum value of **B**. It is called generalized cross validation and will be discussed in Chapter 3.

The above minimization problem can be solved using the structure of dynamic programming and Bellman's Principle of Optimality. This leads to defining the minimum value of E for any initial state and starting at any stage n. Thus

$$f_n(\mathbf{c}) = \min_{\mathbf{g}_j} E_n(\mathbf{c},\mathbf{g}) \tag{1.5-5}$$

Applying the principle of optimality leads to the following recurrence formula

$$\begin{aligned} f_{n-1}(\mathbf{c}) = \min_{\mathbf{g}_{n-1}} \Big[& \big(\mathbf{Q}\mathbf{c} - \mathbf{d}_{n-1}, \mathbf{A}(\mathbf{Q}\mathbf{c} - \mathbf{d}_{n-1})\big) \\ & + (\mathbf{g}_{n-1}, \mathbf{B}\mathbf{g}_{n-1}) + f_n(\mathbf{M}\mathbf{c} + \mathbf{P}\mathbf{g}_{n-1}) \Big] \end{aligned} \tag{1.5-6}$$

This equation represents the classic dynamic programming structure in that the minimum at any point is determined by selecting the decision $\mathbf{g}_{n-1}$ to minimize the immediate cost (first and second terms) and the remaining cost resulting from the decision (the third term). It is important to notice that the minimization is performed over a previously determined optimal function f_n. This structure has succeeded in solving a global optimization problem by replacing it with a series of smaller optimization problems, with the smaller ones being easier to solve.

Equation 1.5-6 represents *two* functions, the optimal cost f_n and the optimal forcing terms $\mathbf{g}_n$. The solution is obtained by starting at the end of the process, n = N, and working backward to n = 1. At the end point, the minimum is determined from

$$f_N(\mathbf{c}) = \min_{\mathbf{g}_N}\left[\left(\mathbf{Qc} - \mathbf{d}_N, \mathbf{A}\left(\mathbf{Qc} - \mathbf{d}_N\right)\right) + \left(\mathbf{g}_N, \mathbf{B}\mathbf{g}_N\right)\right] \quad (1.5\text{-}7)$$

At this end point the minimum is obtained by choosing $\mathbf{g}_N = 0$ which gives

$$f_N(\mathbf{c}) = \left(\mathbf{Qc} - \mathbf{d}_N, \mathbf{A}\left(\mathbf{Qc} - \mathbf{d}_N\right)\right) \quad (1.5\text{-}8)$$

Again, notice that Equation 1.5-7 determines two functions — $\mathbf{g}_N$ and then in turn f_N.

Equation 1.5-8 can be expanded to give

$$f_N(\mathbf{c}) = \left(\mathbf{c}, \mathbf{Q}^T\mathbf{AQc}\right) - 2\left(\mathbf{c}, \mathbf{Q}^T\mathbf{A}\mathbf{d}_N\right) + \left(\mathbf{d}_N, \mathbf{A}\mathbf{d}_N\right) \quad (1.5\text{-}9)$$

The superscript T denotes the transpose of a matrix. The above expression assumes that the matrix $\mathbf{A}$ was symmetric so that $\mathbf{A} = \mathbf{A}^T$. Equation 1.5-9 shows that f_N is quadratic in $\mathbf{c}$. It can be proven inductively that all of the f_n are quadratic in $\mathbf{c}$, so that for any n we can write

$$f_n(\mathbf{c}) = \left(\mathbf{c}, \mathbf{R}_n\mathbf{c}\right) + \left(\mathbf{c}, \mathbf{s}_n\right) + q_n \quad (1.5\text{-}10)$$

where

$\mathbf{R}_n$ is an (m × m) symmetric matrix
$\mathbf{s}_n$ is an (m × 1) vector
q_n is a scalar

The objective now is to obtain recurrence formulas for the matrices $\mathbf{R}_n$, the vectors $\mathbf{s}_n$, and the scalars q_n. These equations will be solved backward starting at n = N with the *initial* conditions obtained from Equation 1.5-9, namely

$$\begin{aligned} \mathbf{R}_N &= \mathbf{Q}^T\mathbf{AQ} \\ \mathbf{s}_N &= -2\mathbf{Q}^T\mathbf{A}\mathbf{d}_N \\ q_N &= (\mathbf{d}_N, \mathbf{A}\mathbf{d}_N) \end{aligned} \quad (1.5\text{-}11)$$

Substituting Equation 1.5-10 into Equation 1.5-6 gives

$$(\mathbf{c},\mathbf{R}_{n-1}\mathbf{c})+(\mathbf{c},\mathbf{s}_{n-1})+q_{n-1}=\min_{g_{n-1}}\Big[\big(\mathbf{Qc}-\mathbf{d}_{n-1},\mathbf{A}(\mathbf{Qc}-\mathbf{d}_{n-1})\big)$$

$$+(\mathbf{g}_{n-1},\mathbf{Bg}_{n-1})+\big(\mathbf{Mc}+\mathbf{Pg}_{n-1},\mathbf{R}_n(\mathbf{Mc}+\mathbf{Pg}_{n-1})\big) \quad (1.5\text{-}12)$$

$$+(\mathbf{Mc}+\mathbf{Pg}_{n-1},\mathbf{s}_n)+q_n\Big]$$

The next steps are to perform the minimization indicated and to substitute the expression for g_{n-1} back into Equation 1.5-12. The minimization gives the equation for the optimal forcing term $\mathbf{g}_{n-1}^*$

$$2\mathbf{Bg}_{n-1}^*+\mathbf{P}^T\mathbf{s}_n+2\mathbf{P}^T\mathbf{R}_n\left(\mathbf{Mc}+\mathbf{Pg}_{n-1}^*\right)=0 \quad (1.5\text{-}13)$$

or

$$\left(2\mathbf{B}+2\mathbf{P}^T\mathbf{R}_n\mathbf{P}\right)\mathbf{g}_{n-1}^*=-\mathbf{P}^T\mathbf{s}_n-2\mathbf{P}^T\mathbf{R}_n\mathbf{Mc} \quad (1.5\text{-}14)$$

At this point it is convenient to define some new matrices $\mathbf{D}(m_g \times m_g)$ and $\mathbf{H}(m_g \times m)$ that will simplify the lengthy expression. These are

$$\mathbf{D}_n=\left(2\mathbf{B}+2\mathbf{P}^T\mathbf{R}_n\mathbf{P}\right)^{-1} \quad (1.5\text{-}15)$$

$$\mathbf{H}_n=2\mathbf{P}^T\mathbf{R}_n \quad (1.5\text{-}16)$$

These can also be expressed as

$$\mathbf{D}_n^{-1}=\left(2\mathbf{B}+2\mathbf{P}^T\mathbf{R}_n\mathbf{P}\right) \quad (1.5\text{-}17)$$

$$\mathbf{H}_n^T=2\mathbf{R}_n\mathbf{P} \quad (1.5\text{-}18)$$

Equation 1.5-14 can now be written as

$$\mathbf{g}_{n-1}^*=-\mathbf{D}_n\mathbf{P}^T\mathbf{s}_n-\mathbf{D}_n\mathbf{H}_n\mathbf{Mc} \quad (1.5\text{-}19)$$

Before substituting this expression into Equation 1.5-12 a considerable amount of simplification can be made by taking the inner product of Equation 1.5-13 with $\mathbf{g}^*_{n-1}$. This gives an identity

$$(\mathbf{g}_{n-1}, \mathbf{B}\mathbf{g}_{n-1}) + (\mathbf{g}_{n-1}, \mathbf{P}^T\mathbf{s}_n)/2 + (\mathbf{g}_{n-1}, \mathbf{P}^T\mathbf{R}_n(\mathbf{M}\mathbf{c} + \mathbf{P}\mathbf{g}_{n-1})) = 0 \quad (1.5\text{-}20)$$

which can be used in Equation 1.5-12 to give

$$\begin{aligned}(\mathbf{c}, \mathbf{R}_{n-1}\mathbf{c}) + (\mathbf{c}, \mathbf{s}_{n-1}) + q_{n-1} &= (\mathbf{Q}\mathbf{c} - \mathbf{d}_{n-1}, \mathbf{A}(\mathbf{Q}\mathbf{c} - \mathbf{d}_{n-1})) \\ &+ (\mathbf{M}\mathbf{c}, \mathbf{R}_n(\mathbf{M}\mathbf{c} + \mathbf{P}\mathbf{g}_{n-1})) + (\mathbf{M}\mathbf{c}, \mathbf{s}_n) + (\mathbf{P}\mathbf{g}_{n-1}, \mathbf{s}_n)/2 + q_n\end{aligned} \quad (1.5\text{-}21)$$

Equation 1.5-19 can now be substituted into the above equation, and equating like powers of **c** will yield (after some lengthy rearranging and taking advantage of symmetric matrix properties)

$$\mathbf{R}_{n-1} = \mathbf{Q}^T\mathbf{A}\mathbf{Q} + \mathbf{M}^T(\mathbf{R}_n - \mathbf{H}_n^T\mathbf{D}_n\mathbf{H}_n/2)\mathbf{M} \quad (1.5\text{-}22)$$

$$\mathbf{s}_{n-1} = -2\mathbf{Q}^T\mathbf{A}\mathbf{d}_{n-1} + \mathbf{M}^T(\mathbf{I} - \mathbf{H}_n^T\mathbf{D}_n\mathbf{P}^T)\mathbf{s}_n \quad (1.5\text{-}23)$$

These are recurrence formulas required to determine the optimal solution of Equation 1.5-4. Equation 1.5-22 is a discrete matrix Riccati equation.

The complete sequence of operations is organized as follows:

A. The backward sweep

1. Solve Equations 1.5-22 and 1.5-23 *backward* starting with the initial conditions given by Equation 1.5-11
2. During the backward sweep store the vectors $\mathbf{D}_n\mathbf{P}^T\mathbf{s}_n$ ($m_g \times 1$) and the matrices $\mathbf{D}_n\mathbf{H}_n\mathbf{M}$ ($m_g \times m$) as required by Equation 1.5-19.

B. The forward sweep

1. Start with the initial condition for **x**
2. Compute the optimal $\mathbf{g}_2$ using Equation 1.5-19

3. Compute the next state $\mathbf{x}_2$ using Equation 1.5-1
4. Repeat the steps for $\mathbf{g}_3$, $\mathbf{x}_3$, etc.

1.5.1 Optimal Initial Conditions

In regard to the initial conditions for $\mathbf{x}_1$, two options can be used. Which one applies will depend on the particular problem. In most cases, the initial conditions are known with such a high degree of certainty that they can be used directly. This is usually the case where the number of state variables is much larger than the number of measurements. Recall, however, that the optimum solution is valid for any set of initial conditions. Thus, in some cases it may be advantageous to compute an optimal set of initial conditions. This can be done by using Equation 1.5-10 for n = 1

$$f_1(\mathbf{c}) = (\mathbf{c}, \mathbf{R}_1\mathbf{c}) + (\mathbf{c}, \mathbf{s}_1) + q_1 \tag{1.5.1-1}$$

Recall that this represents the minimum starting with any state $\mathbf{c}$. Thus it follows that the optimal set of initial conditions is that which minimizes Equation 1.5.1-1. This gives

$$\mathbf{c}^* = -(\mathbf{R}_1)^{-1}\mathbf{s}_1/2 \tag{1.5.1-2}$$

This assumes that the inverse of $\mathbf{R}_1$ exists and that $\mathbf{c}^*$ can be calculated.

There are many computational issues that arise in the calculation of the various matrices, especially when matrices of different order are involved. In Chapter 3 we will present a method that avoids the calculation and storage of the matrices $\mathbf{R}_n$. Clearly if we are going to solve systems with 1000 state variables it will be essential to avoid full matrices of the order 1000 × 1000. Even in working with smaller systems where storing the $\mathbf{R}_n$'s is acceptable, one should be on the alert in order to avoid needless calculations. For example, the number of multiplications required to multiply two matrices of the order $(n_a \times n_b)$ and $(n_b \times n_c)$ is $n_a n_b n_c$. Now consider the term $\mathbf{D}_n\mathbf{P}^T\mathbf{s}_n$ where m_g is 5 and m is 100, $\mathbf{D}$ is $(m_g \times m_g)$, $\mathbf{P}$ is $(m \times m_g)$, and $\mathbf{s}$ is $(m \times 1)$. If one first carelessly calculated $\mathbf{D}_n\mathbf{P}^T$ this would require 2500 multiplications plus an additional 500 to complete the calculations for the term $\mathbf{D}_n\mathbf{P}^T\mathbf{s}_n$.

However, if one first calculated $\mathbf{P}^T\mathbf{s}_n$ this would require only 500 multiplications and an additional 25 for completion. This is a difference of 2475 multiplications to get the same result. Of course one should always take advantage of symmetric matrices to store one half of the matrix and to further reduce the multiplications. These simple ideas represent a first level of computational efficiency. There has been a considerable amount of research done to further reduce the number of computations. These are mostly in the areas of Kalman filtering. See the work by Bierman (1977).

1.5.2 Simple Multidimensional System

As an example of the simplest multidimensional system consider a system of order m = 2 that represents a continuous function with a continuous first derivative. In the interval $t_i < t < t_{i+1}$, the function y is given as

$$y(t) = y_i + v_i(t - t_i) + a_i(t - t_i)^2/2 \tag{1.5.2-1}$$

and the derivative as

$$y'(t) = v_i + a_i(t - t_i) \tag{1.5.2-2}$$

For a constant interval $h = t_{i+1} - t_i$, the above equations can be represented as a discrete system

$$\begin{bmatrix} y_{i+1} \\ v_{i+1} \end{bmatrix} = \begin{bmatrix} 1 & h \\ 0 & 1 \end{bmatrix} \begin{bmatrix} y_i \\ v_i \end{bmatrix} + \begin{bmatrix} h^2/2 \\ h \end{bmatrix} [a_i] \tag{1.5.2-3}$$

The forcing term is represented by a_i and is constant in the interval (t_i, t_{i+1}). Thus, for this system, the matrices **M** and **P** of Equation 1.5-1 become

$$\mathbf{M} = \begin{bmatrix} 1 & h \\ 0 & 1 \end{bmatrix}; \quad \mathbf{P} = \begin{bmatrix} h^2/2 \\ h \end{bmatrix} \tag{1.5.2-4}$$

For the data we will use the same figures in Section 1.4 and listed in Table 1.1. The timestep h is 0.10 for this example. Since we are comparing the data d_i with y_i, the matrix **Q** of Equation 1.5-3 becomes a matrix of order (1 × 2)

$$\mathbf{Q} = \begin{bmatrix} 1 & 0 \end{bmatrix} \tag{1.5.2-5}$$

so that

$$\mathbf{z}_i = \begin{bmatrix} 1 & 0 \end{bmatrix} \begin{bmatrix} y_i \\ v_i \end{bmatrix}$$

The forcing vector g_i becomes a_i. The matrices **A** and **B** of Equation 1.5-4 are of order (1 × 1). **A** will be set to 1.0 and **B** to the scalar b.

In this example there are 11 steps, and the backward sweep begins with the initial conditions given by Equation 1.5-11. These are

$$\mathbf{R}_{11} = \begin{bmatrix} 1 \\ 0 \end{bmatrix} [1] \begin{bmatrix} 1 & 0 \end{bmatrix} = \begin{bmatrix} 1 & 0 \\ 0 & 0 \end{bmatrix}$$

and

$$\mathbf{s} = -2 \begin{bmatrix} 1 \\ 0 \end{bmatrix} [1][1.9701] = \begin{bmatrix} -3.940 \\ 0 \end{bmatrix}$$

Since h = 0.10, the matrices **M** and **P** become

$$\mathbf{M} = \begin{bmatrix} 1 & 0.10 \\ 0 & 1 \end{bmatrix}; \quad \mathbf{P} = \begin{bmatrix} 0.005 \\ 0.10 \end{bmatrix} \tag{1.5.2-6}$$

and $\mathbf{H}_{11}$ (Equation 1.5-16) becomes a 1 × 2 matrix

$$\mathbf{H}_{11} = 2\begin{bmatrix} 0.005 & 0.10 \end{bmatrix} \begin{bmatrix} 1 & 0 \\ 0 & 0 \end{bmatrix} = \begin{bmatrix} 0.010 & 0.0 \end{bmatrix}$$

and $\mathbf{D}_{11}$ (Equation 1.5-17) becomes a scalar (1 × 1)

$$\mathbf{D}_{11}^{-1} = 2b + 2\begin{bmatrix} 0.005 & 0.10 \end{bmatrix}\begin{bmatrix} 1 & 0 \\ 0 & 0 \end{bmatrix}\begin{bmatrix} 0.005 \\ 0.10 \end{bmatrix}$$

or

$$\mathbf{D}_{11}^{-1} = 2b + 0.00005$$

In this example, we will set the regularization parameter b equal to 1.0E–05 so $\mathbf{D}_{11}^{-1} = 0.00007$ or $\mathbf{D}_{11} = 14286.0$. With these values, the matrix $\mathbf{R}_{10}$ can be calculated using Equation 1.5-22. Carrying out the calculations gives

$$\mathbf{R}_{10} = \begin{bmatrix} 1.2857 & 0.02857 \\ 0.02857 & 0.002857 \end{bmatrix}$$

Similarly, Equation 1.5-23 gives

$$\mathbf{s}_{10} = \begin{bmatrix} -5.1457 \\ -0.1126 \end{bmatrix}$$

The recurrence equations for $\mathbf{R}_n$ and $\mathbf{s}_n$ are continued until the final step is reached. The final matrix $\mathbf{R}_1$ is

$$\mathbf{R}_1 = \begin{bmatrix} 1.439 & 0.03162 \\ 0.03162 & 0.002971 \end{bmatrix}$$

It becomes apparent that even the simplest multidimensional case requires a computer program and if one is written, it might as well be written for the general multidimensional case.

During the backward calculations the vectors $\mathbf{D}_n\mathbf{P}^T\mathbf{s}_n$ and the matrices $\mathbf{D}_n\mathbf{H}_n\mathbf{M}$ as required by Equation 1.5-19 are saved for each step. The initial conditions are determined using Equation 1.5.1-2. The forward sweep can now be completed, and the final results are shown in Table 1.3. The v_i have been plotted in Figure 1.4 and compared against the noiseless original smooth curve. Compare this result with those of the first-order system in Figure 1.3. The higher order system has yielded smoother estimates.

TABLE 1.3

Results of Forward Sweep

STEP	d_i	y_i	v_i	a_i
1	–1.000E–02	–1.086E–02	8.312E–01	4.291E–01
2	9.000E–02	7.441E–02	8.741E–01	9.083E+00
3	1.800E–01	2.072E–01	1.782E+00	1.191E+01
4	4.400E–01	4.450E–01	2.974E+00	–1.394E+00
5	7.600E–01	7.354E–01	2.834E+00	–4.945E+00
6	1.000E+00	9.941E–01	2.340E+00	6.709E+00
7	1.220E+00	1.262E+00	3.011E+00	4.564E–01
8	1.600E+00	1.565E+00	3.056E+00	–9.134E+00
9	1.800E+00	1.825E+00	2.143E+00	–1.371E+01
10	2.010E+00	1.971E+00	7.716E–01	–1.112E+01
11	1.970E+00	1.993E+00	–3.406E–01	0.0

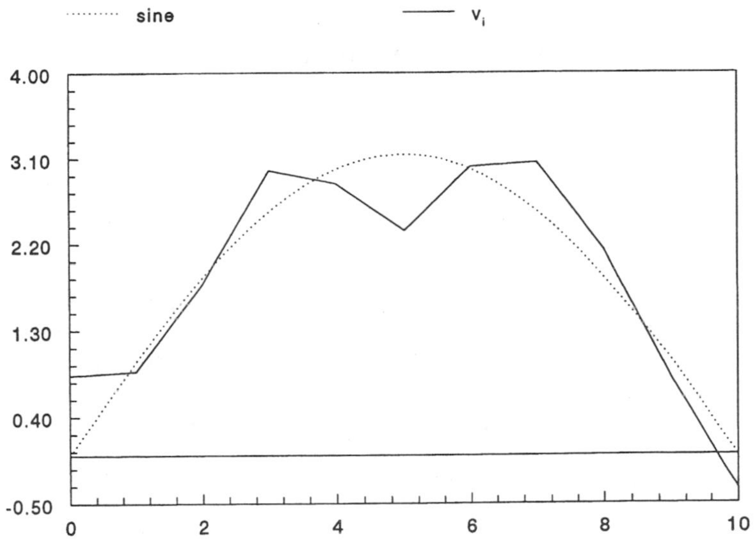

FIGURE 1.4
Plot of v_i for b = 1.E–05

In a later chapter these data will be used with a third-order filter, which would make the matrices 3 × 3. The sequence of the computational steps is exactly the same.

1.5.3 Principle of Superposition

There is one other general multidimensional system that occurs very frequently. This is the case of a system with an additional *known* forcing term.

$$\mathbf{x}_{j+1} = \mathbf{M}\mathbf{x}_j + \mathbf{v}_j + \mathbf{P}\mathbf{g}_j \tag{1.5.3-1}$$

where **v** is an (m × 1) vector representing the known forcing term. This situation will also arise in the analysis of nonlinear systems.

This case can be analyzed using the *Principle of Superposition.* First, let $\mathbf{x}_j$ be a linear combination of two other vectors $\mathbf{y}_j$ and w_j

$$\mathbf{x}_j = \mathbf{y}_j + \mathbf{w}_j \tag{1.5.3-2}$$

Substituting this into Equation 1.5.3-1 gives

$$\mathbf{y}_{j+1} + \mathbf{w}_{j+1} = \mathbf{M}\mathbf{y}_j + \mathbf{M}\mathbf{w}_j + \mathbf{v}_j + \mathbf{P}\mathbf{g}_j \tag{1.5.3-3}$$

We can now separate the equation into two systems

$$\mathbf{y}_{j+1} = \mathbf{M}\mathbf{y}_j + \mathbf{P}\mathbf{g}_j \tag{1.5.3-4}$$

$$\mathbf{w}_{j+1} = \mathbf{M}\mathbf{w}_j + \mathbf{v}_j \tag{1.5.3-5}$$

The equation for $\mathbf{w}_j$ will always use the initial conditions $\mathbf{w}_j = 0$ and can be solved directly for all of the $\mathbf{w}_j$'s. The variable $\mathbf{z}_j$, which relates to the data, must now be adjusted to incorporate the $\mathbf{w}_j$'s. Equation 1.5-3 becomes

$$\mathbf{z}_j = \mathbf{Q}\mathbf{x}_j = \mathbf{Q}\mathbf{y}_j + \mathbf{Q}\mathbf{w}_j \tag{1.5.3-6}$$

and the general least squares error Equation 1.5-4 is rewritten as

$$E(\mathbf{c}, \mathbf{g}_j) = \left(\mathbf{Q}\mathbf{y}_j + \mathbf{Q}\mathbf{w}_j - \mathbf{d}_j, \mathbf{A}\left(\mathbf{Q}\mathbf{y}_j + \mathbf{Q}\mathbf{w}_j - \mathbf{d}_j\right)\right) + \left(\mathbf{g}_j, \mathbf{B}\mathbf{g}_j\right) \tag{1.5.3-7}$$

If we define a new set of data $\mathbf{d}_j^*$ as

$$\mathbf{d}_j^* = \mathbf{d}_j - \mathbf{Q}\mathbf{w}_j \tag{1.5.3-8}$$

then Equation 1.5.3-7 becomes

$$E\left(c,g_j\right)=\left(Qy_j-d^*_{jj},A\left(Qy_j-d^*_j\right)\right)+\left(g_j,Bg_j\right)$$

Thus the optimization problem involves only $\mathbf{y}_j$'s and is solved in exactly the same manner as before, using the data $\mathbf{d}^*_j$. The final solution is obtained by adding $\mathbf{y}_j$ and $\mathbf{w}_j$.

In the analysis of a nonlinear system, the vectors $\mathbf{v}_j$ will represent nonlinear terms that are obtained by linearizing the equations using a constant matrix **M**. This will introduce an additional global iteration loop that will involve updating the vectors $\mathbf{v}_j$ to reflect the current optimal solution. This will be discussed in further detail in Chapter 7.

Exercise

Rederive the general optimization problem for the case where **A** and **B** also vary with the timestep. That is, consider $\mathbf{A}_j$ and $\mathbf{B}_j$ in Equation 1.5-4 to be functions of time (j).

1.6 OPTIMAL CONTROL AS A MULTISTAGE DECISION PROCESS

The proceeding section can also be viewed as a special case of an optimal control process. It is referred to as the linear regulator with a quadratic criterion. The criterion, as expressed by Equation 1.5-4, is intended to match a given trajectory (the first term) while using a finite amount of control (the second term). The demarcation between our applications and a true control process is that a control process takes place in real time, while most of our applications are based on the assumption that we have the entire set of measurements at our disposal instead of obtaining them one at a time and having to make instantaneous decisions. These two different points of view lead to different emphasis on computations, yet the underlying mathematics is very similar. The controller and its forces on the system can be viewed as representing an unknown function that has exerted a force on the system. The reader can review Bellman (1967), Lapidus and Luus (1967), and Bryson and Ho (1975) for a sampling of the voluminous material written on optimal control. There are also several other types of control problems, such as terminal control and minimum time, that fit into the same general methodology.

A common situation in optimal control theory is the case of designing a controller to keep a system near an equilibrium state. In this case the system equations are rewritten to reflect a zero equilibrium state. This can be represented in the derivation of the previous section by setting all of the d_j's to zero. The criterion now becomes

$$E(\mathbf{c},\mathbf{g}_j) = \sum_{j=1}^{N} \left(\left(\mathbf{x}_j, \mathbf{A}\mathbf{x}_j \right) + \left(\mathbf{g}_j, \mathbf{B}\mathbf{g}_j \right) \right) \tag{1.6-1}$$

Without the data $\mathbf{d}_j$ the vectors $\mathbf{s}_n$ are all zero and do not enter into the solution. The equations for the $\mathbf{R}_n$ remain unchanged. An additional step in the design of a control system is to consider the asymptotic behavior of $\mathbf{R}_n$. If it is further assumed that the $\mathbf{R}_n$ reach a steady-state $\mathbf{R}$ then the steady-state optimal controller becomes

$$\mathbf{g} = -\left(2\mathbf{B} + 2\mathbf{P}^T\mathbf{R}\mathbf{P}\right)^{-1} 2\mathbf{P}^T\mathbf{R}\mathbf{M}\mathbf{c} \tag{1.6-2}$$

which is precisely a feedback controller. That is, the forces $\mathbf{g}$ are proportional to the current state of the system $\mathbf{c}$. It has tacitly been assumed that the system will be constantly perturbed by unknown (usually random) external forces so that the state of the system will change at each step.

A reasonable computational method for computing the steady-state $\mathbf{R}$ is to compute the sequence $\mathbf{R}_n$ and monitor it until it does not change (or remains within some small tolerance). A suitable matrix norm can be used to measure the convergence.

A close inspection of the optimal controller reveals that it requires a knowledge of the entire state $\mathbf{c}$ in real time to compute the controlling forces. Usually it is rare that all of the state variables are measured or even measurable. This leads to another broad area known as *optimal estimation* which includes Kalman filtering. Its purpose is to optimally estimate the current state $\mathbf{c}$

from measurements being taken, again, in real time. It is no surprise that all of these areas are mathematically related. In a later chapter we will discuss the optimal estimation problem in relation to a system identification problem. The reader can find further information on optimal control in the references. We shall not pursue this subject, with all of the variations involved in actually implementing such controllers. The main difficulty is that one must control a process taking place in real time. In our approach we will assume that the measurements have been taken off line and there is no real time constraint.

As a simple example of a steady controller consider the first-order example of Section 1.4. The steady-state equation for r is given by Equation 1.4-16.

$$r = 1 + br/(b + r)$$

which for b = 0.2 can be solved to give r = 1.171. An inspection of Table 1.1 shows how quickly the steady-state value was reached by using the recurrence formula. Subsequently the steady-state control force g is given by Equation 1.4-15

$$g^* = -2rc/(2b + 2r) = -0.854c$$

This is coupled with the system Equation 1.4-1 to give the final system

$$x_{j+1} = x_j + g^* = 0.146x_j$$

It is clear that the original system has now been altered by the control force so that it will remain near zero. Without the controller, the system was neutrally stable.

Exercise

Consider the unstable system where m = 2 in Equation 1.4-1. Compute the controller and show that the resulting system is now stable.

As another example of a regulator, consider the second-order system of Section 1.5. The system equation is given by

$$\begin{bmatrix} y_{i+1} \\ v_{i+1} \end{bmatrix} = \begin{bmatrix} 1 & h \\ 0 & 1 \end{bmatrix} \begin{bmatrix} y_i \\ v_i \end{bmatrix} + \begin{bmatrix} h^2/2 \\ h \end{bmatrix} [a_i] \tag{1.6-3}$$

For the value of $b = 1.0E{-5}$ and $h = 0.1$ the steady-state matrix R was calculated and used in Equation 1.6-2 to give the controller (g_i is a_i). In this case there are two state variables so that

$$a = \begin{bmatrix} -96.532 & -13.895 \end{bmatrix} \begin{bmatrix} y_i \\ v_i \end{bmatrix}$$

Substituting this into Equation 1.5.2-3 gives the complete system.

$$\begin{bmatrix} y_{i+1} \\ v_{i+1} \end{bmatrix} = \begin{bmatrix} 1 & 0.1 \\ 0 & 1 \end{bmatrix} \begin{bmatrix} y_i \\ v_i \end{bmatrix} + \begin{bmatrix} 0.005 \\ 0.10 \end{bmatrix} \begin{bmatrix} -96.532 & -13.895 \end{bmatrix} \begin{bmatrix} y_i \\ v_i \end{bmatrix} \tag{1.6-4}$$

completing the operations gives

$$\begin{bmatrix} y_{i+1} \\ v_{i+1} \end{bmatrix} = \begin{bmatrix} 0.5173 & 0.0305 \\ -9.653 & -0.3895 \end{bmatrix} \begin{bmatrix} y_i \\ v_i \end{bmatrix} \tag{1.6-5}$$

which is now a stable system. Notice that the criteria only required that y_i remain small; v_i was not included in the criteria as defined by matrix **Q**. If one wanted both y_i and v_i to be controlled, then **Q** (Equation 1.5-3) would be redefined as the identity matrix **I** (2×2).

Exercise

Compute the controller for **Q** = **I**.

The following exercises illustrate several interesting applications of dynamic programming.

Exercise — Optimal Triangulation

Given the location of the vertices of a polygon, the problem is to divide the polygon into triangles. The total length of the chords is to be minimized. No two chords can intersect (see Fuchs, 1971).

Exercise — Jeep Crossing

A jeep is to cross a desert 200 miles long, but it can only carry enough gasoline to travel 100 miles. The idea is to leave caches of gasoline which can be picked up later. Set the distance between caches to 20 miles. The problem is to cross with a minimum of gasoline (see Fine, 1947).

Exercise — Optimal Binary Search Tree

A binary search tree can be constructed for n variables that are ordered: $x_1 < x_2 < \ldots < x_n$. Suppose that the probability of requesting any x_i is given as p_i. Construct a binary tree that minimizes the searching. The cost of a search is given as the sum of $p_i(1 + \text{level}(x_i))$ (see Knuth, 1971).

Exercise — Optimal Piecewise Curve Fitting

The fitting of a set of piecewise linear curves to a set of data is fairly easy if one knows the breakpoints. The difficult problem is to determine the optimal spacing of the breakpoints such that the total least squares error is minimized (see Bellman, Vol. 2, 1967).

Exercise — Routing Problem

In the problem in Section 1.2, the nodes were only connected to the immediate nodes and the choice of travel was limited either to the right or up. In the general routing problem each node can be connected to all of the others. Let t_{ij} be the cost to travel between any two points i and j and let

$$f_i = \text{minimum cost to go from i to N}; \quad f_N = 0$$

Then the principle of optimality yields

$$f_i = \min_{j \neq i} \left[t_{ij} + f_j \right] \qquad i = 1, 2, 3, \ldots, N-1$$

This problem requires iterative solutions (see Bellman, Vol. 2, 1967).

CHAPTER 2

MATRICES AND DIFFERENTIAL EQUATIONS

2.1 INTRODUCTION

In this chapter we shall be concerned with differential equations and the general use of vector matrix notation. Only a brief review of pertinent information needed throughout the book will be given in this chapter. It is assumed that the reader is familiar with the subject in general.

2.2 VECTOR-MATRIX CALCULUS

Let n be a given positive integer. A system of n complex numbers $x_1, x_2, x_3, \ldots, x_n$ is called an *n dimensional vector* and, when written in the form

$$\mathbf{x} = \begin{bmatrix} x_1 \\ x_2 \\ x_3 \\ \vdots \\ x_n \end{bmatrix} \tag{2.2-1}$$

is called a *column vector*. If the vector is written as

$$\mathbf{x} = \begin{pmatrix} x_1 & x_2 & x_3 & \ldots & x_n \end{pmatrix} \tag{2.2-2}$$

it is called a *row vector*.

In general, boldface letters **x,y,z** or **a,b,c** will be used to denote vectors throughout the text. The elements x_i are called the components of the vector **x**. The length of the vector is the number of components n.

If **x** and **y** are two vectors, then **x** is equal to **y**, that is, **x** = **y**, if and only if the n equalities

$$x_1 = y_1, \quad x_2 = y_2, \quad x_3 = y_3, \quad \ldots, \quad x_n = y_n \tag{2.2-3}$$

are satisfied.

The sum of two vectors **x** and **y** is defined to be the vector **z** and is written

$$\mathbf{z} = \mathbf{x} + \mathbf{y} = \begin{bmatrix} x_1 + y_1 \\ x_2 + y_2 \\ x_3 + y_3 \\ \vdots \\ x_n + y_n \end{bmatrix} \tag{2.2-4}$$

It is clear that addition is commutative and associative. We define subtraction in a similar manner.

Another operation is that of multiplying a vector **x** by a scalar a. The multiplication is defined by

$$\mathbf{z} = a\mathbf{x} = \mathbf{x}a = \begin{bmatrix} ax_1 \\ ax_2 \\ ax_3 \\ \vdots \\ ax_n \end{bmatrix} \tag{2.2-5}$$

The operations of addition and multiplication given in Equation 2.2-4 and 2.2-5 always yield elements that belong to a vector **z**. Another useful operation combines two elements in such a way that we obtain a real number instead of a vector. This operation is known as the *inner product*. If **x** and **y** are any two vectors, then the inner product of **x** and **y** [*denoted* (**x**,**y**)] is defined by

$$(\mathbf{x},\mathbf{y}) = \sum_{i=1}^{N} x_i y_i \qquad (2.2\text{-}6)$$

It is noted that the inner product is a scalar and is symmetric, that is

$$(\mathbf{x},\mathbf{y}) = (\mathbf{y},\mathbf{x}) \qquad (2.2\text{-}7)$$

When we combine the definition of inner product with the definitions of vector addition and scalar multiplication, we obtain the following:

$$(a\mathbf{x},\mathbf{y}) = a(\mathbf{x},\mathbf{y}) = (\mathbf{x},a\mathbf{y}) \qquad (2.2\text{-}8a)$$

$$(\mathbf{x}+\mathbf{y},\mathbf{r}+\mathbf{s}) = (\mathbf{x},\mathbf{r}) + (\mathbf{x},\mathbf{s}) + (\mathbf{y},\mathbf{r}) + (\mathbf{y},\mathbf{s}) \qquad (2.2\text{-}8b)$$

If $(\mathbf{x},\mathbf{y}) = 0$ then the two vectors **x** and **y** are said to be *orthogonal.*

If complex components are used, the inner product is written as

$$(\mathbf{x},\mathbf{y}) = \sum_{i=1}^{N} x_i y_i^* \qquad (2.2\text{-}9)$$

where y_i^* is the complex conjugate of y_i. Note however, that for this case the inner product is no longer symmetric; that is

$$(\mathbf{x},\mathbf{y}) = (\mathbf{y},\mathbf{x})^* \qquad (2.2\text{-}10)$$

If all the components of a vector are real, we can form the inner product as

$$(\mathbf{x},\mathbf{x}) = x_i^2 \qquad (2.2\text{-}11)$$

which represents the Euclidean length of the vector **x**. A convenient notation often used to denote the length of a vector **x** is

$$\|x\| = \left(\sum_{i=1}^{N} x_i^2 \right)^{1/2} \tag{2.2-12}$$

and is called the *norm of the vector* **x**.

A matrix can be described as a rectangular array of elements that can be written in the form

$$\mathbf{A} = \begin{bmatrix} a_{11} & a_{12} & a_{13} & \cdots & a_{1n} \\ a_{21} & a_{22} & a_{23} & \cdots & a_{2n} \\ \vdots & \vdots & \vdots & \vdots & \vdots \\ a_{m1} & a_{m2} & a_{m3} & \cdots & a_{mn} \end{bmatrix} \tag{2.2-13}$$

consisting of m rows and n columns. Matrices will be represented by bold upper case letters **A**, **B**, **C**, etc. The shorthand notation is given by

$$\mathbf{A} = \left[a_{ij} \right] \tag{2.2-14}$$

and will be used to conserve space. Two matrices $\mathbf{A} = [a_{ij}]$, and $\mathbf{B} = [b_{ij}]$ are equal if and only if $a_{ij} = b_{ij}$ for all values of i and j.

The matrix which is obtained from $\mathbf{A} = [a_{ij}]$ by interchanging rows and columns is called the *transpose of matrix* **A**, and is written as

$$\mathbf{A}^T = \begin{bmatrix} a_{11} & a_{21} & a_{31} & \cdots & a_{m1} \\ a_{12} & a_{22} & a_{32} & \cdots & a_{m2} \\ \vdots & \vdots & \vdots & \vdots & \vdots \\ a_{1n} & a_{2n} & a_{3n} & \cdots & a_{mn} \end{bmatrix} \tag{2.2-15}$$

where the symbol T denotes the transpose. Thus, if the element of row i and column j of **A** is a_{ij} then the element in row i and column j of $\mathbf{A}^T$ is a_{ji}.

A matrix that has the number of rows equal to the number of columns is called a *square matrix*. Thus, a square matrix of order 3 is written

$$\mathbf{A} = \begin{bmatrix} a_{11} & a_{12} & a_{13} \\ a_{21} & a_{22} & a_{23} \\ a_{32} & a_{32} & a_{33} \end{bmatrix} \tag{2.2-16}$$

For this example, the elements a_{11}, a_{22}, and a_{33} lie on and form the main or principal diagonal. A square matrix for which the elements below the principal diagonal are zero is called an *upper triangular matrix*. A square matrix for which the elements above the principal diagonal are zero is called a *lower triangular matrix*.

A matrix that has every element equal to zero is called a *null* or *zero matrix* and is denoted by the symbol **0**.

A square matrix is called a *unit* or *identity matrix* if all the elements on the principal diagonal are unity and the remaining elements are zero. For a square matrix of order 3, the unit or identity matrix has the form

$$\mathbf{I} = \begin{bmatrix} 1 & 0 & 0 \\ 0 & 1 & 0 \\ 0 & 0 & 1 \end{bmatrix} \tag{2.2-17}$$

If all the elements of a square matrix other than those along the main diagonal are zero, the matrix is called a *diagonal matrix* and, for a matrix of order 3, is written

$$\mathbf{A} = \begin{bmatrix} a_{11} & 0 & 0 \\ 0 & a_{22} & 0 \\ 0 & 0 & a_{33} \end{bmatrix} \tag{2.2-18}$$

A square matrix is said to be *symmetric* if it is equal to its transpose, that is $\mathbf{A} = \mathbf{A}^T$ or $a_{ij} = a_{ji}$ and *antisymmetric* or *skew-symmetric* if it is equal to the negative of its transpose, that is $\mathbf{A} = -\mathbf{A}^T$ or $a_{ij} = -a_{ji}$.

If s is a scalar, the multiplication of a matrix by a scalar is defined as

$$s\mathbf{A} = \mathbf{A}s = \left[sa_{ij}\right] \tag{2.2-19}$$

The multiplication of a vector **x** by a matrix **A** is the vector **y** (**y** = (y_i), i = 1,2,...,n) such that

$$y_i = (\mathbf{Ax})_i = a_{ij}x_j \qquad (2.2\text{-}20)$$

or in matrix notation

$$\mathbf{y} = \mathbf{Ax} \qquad (2.2\text{-}21)$$

For the multiplication of two matrices **A** and **B**, we can consider another transformation given by

$$\mathbf{z} = \mathbf{By}$$

Then using Equation 2.2-21 we have

$$\mathbf{z} = \mathbf{By} = \mathbf{B}(\mathbf{Ax})$$

or

$$\mathbf{z} = (\mathbf{BA})\mathbf{x}$$

Thus, the ij_{th} element of **BA** is c_{ij} such that

$$c_{ij} = (\mathbf{BA})_{ij} = b_{ik}a_{ki} \qquad (2.2\text{-}22)$$

or in standard matrix notation

$$\mathbf{C} = \mathbf{BA} \qquad (2.2\text{-}23)$$

In general, matrix multiplication is not commutative. That is,

$$\mathbf{BA} = \mathbf{AB}$$

However, associativity of multiplication does hold, so that

$$(\mathbf{AB})\mathbf{C} = \mathbf{A}(\mathbf{BC})$$

We have previously defined the inner product of two vectors **x** and **y**. We now define the inner product of **y** and **Ax**, that is

$$(\mathbf{y}, \mathbf{Ax}) = y_i a_{ij} x_j$$

and it is easily shown that

$$(\mathbf{y}, \mathbf{Ax}) = (\mathbf{A}^T\mathbf{y}, \mathbf{x}) \tag{2.2-24}$$

If we consider **A** to be a real n × n symmetric matrix such that the quadratic form (Bellman, 1960b)

$$(\mathbf{x}, \mathbf{Ax}) = a_{ij} x_i x_j$$

is positive for all real vectors $\mathbf{x} \neq \mathbf{0}$, then **A** is said to be *positive definite* and hence

$$\mathbf{A} > 0 \tag{2.2-25}$$

If $(\mathbf{x},\mathbf{Ax}) \geq \mathbf{0}$ for all real vectors $\mathbf{x} \neq \mathbf{0}$, then **A** is said to be *nonnegative definite* and hence

$$\mathbf{A} \geq 0 \tag{2.2-26}$$

The inverse of a matrix which is denoted as

$$\mathbf{A}^{-1} = \text{Inverse of } \mathbf{A} \tag{2.2-27}$$

exists only if the determinant of **A** does not vanish. It has the property

$$\mathbf{AA}^{-1} = \mathbf{A}^{-1}\mathbf{A} = \mathbf{I} \tag{2.2-28}$$

If $\mathbf{Ax} = \mathbf{y}$, then $\mathbf{x} = \mathbf{A}^{-1}\mathbf{y}$.

For the two-dimensional case we have

$$a_{11}x_1 + a_{12}x_2 = y_1$$

$$a_{21}x_1 + a_{22}x_2 = y_2$$

Solving for x_1 and x_2 yields

$$\begin{bmatrix} x_1 \\ x_2 \end{bmatrix} = \frac{1}{|\mathbf{A}|} \begin{bmatrix} a_{22} & -a_{12} \\ -a_{21} & a_{11} \end{bmatrix} \begin{bmatrix} y_1 \\ y_2 \end{bmatrix} \quad (2.2\text{-}29)$$

which can be generalized to

$$\left(\mathbf{A}^{-1}\right)_{ij} = \text{cofactor of } A_{ji} / |\mathbf{A}| \quad (2.2\text{-}30)$$

where $|\mathbf{A}|$ is the determinant of **A**.

If **A** and **B** are both nonsingular, then their product is nonsingular and

$$(\mathbf{AB})^{-1} = \mathbf{B}^{-1}\mathbf{A}^{-1} \quad (2.2\text{-}31)$$

Also, we have

$$\left(\mathbf{A}^{-1}\right)^{-1} = \mathbf{A} \quad (2.2\text{-}32)$$

Another matrix inversion formula that is important in control theory and used throughout the book is the *Sherman Morrison equation* which is given as

$$(\mathbf{A} + \mathbf{BDC})^{-1} = \mathbf{A}^{-1} - \mathbf{A}^{-1}\mathbf{B}\left(\mathbf{D}^{-1} + \mathbf{CA}^{-1}\mathbf{B}\right)^{-1}\mathbf{CA}^{-1} \quad (2.2\text{-}33)$$

2.2.1 Eigenvectors and Eigenvalues

Consider an $n \times n$ matrix **A** given by

$$\mathbf{A} = \begin{bmatrix} a_{11} & a_{12} & \cdots & a_{1n} \\ a_{21} & a_{22} & \cdots & a_{2n} \\ \vdots & \vdots & \vdots & \vdots \\ a_{n1} & a_{n2} & \cdots & a_{nn} \end{bmatrix}$$

and introduce a column matrix **u** which is (n × 1). If

$$\mathbf{Au}^{(i)} = L_i \mathbf{u}^{(i)} \qquad i = 1, 2, \ldots, n \tag{2.2.1-1}$$

then $\mathbf{u}^{(i)}$ is called the *eigenvector* and L_i the *eigenvalue* of matrix **A**, a number, possibly complex. There are n eigenvalues where

$$|\mathbf{A} - \mathbf{LI}| = 0 \tag{2.2.1-2}$$

gives the eigenvalues. Equation 2.2.1-2 is also called the characteristic equation. Equation (2.2.1-1), becomes, when written in full,

$$\begin{aligned} a_{11}u_1 + a_{12}u_2 + \ldots + a_{1n}u_n &= Lu_1 \\ a_{21}u_1 + a_{22}u_2 + \ldots + a_{2n}u_n &= Lu_2 \\ &\cdots\cdots \\ a_{n1}u_1 + a_{n2}u_2 + \ldots + a_{nn}u_n &= Lu_n \end{aligned} \tag{2.2.1-3}$$

A nontrivial solution of Equation 2.2.1-3 is possible if and only if

$$\begin{vmatrix} a_{11} - L & a_{12} & \cdots & a_{1n} \\ a_{21} & a_{22} - L & \cdots & a_{2n} \\ \vdots & \vdots & \vdots & \vdots \\ a_{n1} & a_{n2} & \cdots & a_{nn} - L \end{vmatrix} = 0 \quad \text{or} \quad |\mathbf{A} - \mathbf{LI}| = 0 \tag{2.2.1-4}$$

When we expand this determinant, we obtain an algebraic equation in L which can be written as

$$L^n + c_1 L^{n-1} + c_2 L^{n-2} + \ldots + c_n = 0 \tag{2.2.1-5}$$

Equation 2.2.1-5 is the characteristic equation of the matrix **A**. As an example, consider the equations

$$4x_1 - 2x_2 = Lx_1$$

$$x_1 + x_2 = Lx_2$$

The corresponding characteristic equation is given by expanding the determinant

$$\begin{vmatrix} 4-L & -2 \\ 1 & 1-L \end{vmatrix} = 0$$

which yields

$$L^2 - 5L + 6 = 0$$

The roots which are the eigenvalues are found to be $L_1 = 3$ and $L_2 = 2$.

To obtain the eigenvector we substitute back into our original equation the corresponding eigenvalues. For $L_1 = 3$ we find

$$x_1 - 2x_2 = 0$$

$$x_1 - 2x_2 = 0$$

Assuming $x_1 = 2$, we find $x_2 = 1$. For $L_2 = 2$, our equations become

$$2x_1 - 2x_2 = 0$$

$$x_1 - x_2 = 0$$

Again, assuming $x_1 = 1$, we find $x_2 = 1$. Thus, the eigenvectors $\mathbf{x}_1^T = (2\ 1)$, and $\mathbf{x}_2^T = (1\ 1)$ correspond to the determined eigenvalues $L_1 = 3$ and $L_2 = 2$.

A function f(A) of an $n \times n$ matrix **A** can be replaced by a polynomial p(**A**) in **A** of order n – 1 (Hohn, 1957).

$$f(\mathbf{A}) = p(\mathbf{A}) = c_0\mathbf{I} + c_1\mathbf{A} + c_2\mathbf{A}^2 + \ldots + c_{n-1}\mathbf{A}^{n-1} \quad (2.2.1\text{-}6)$$

The constants $c_0, c_1, c_2, \ldots, c_n$ can be determined by using the fact that the eigenvalues L_i also satisfy Equation 2.2.1-6, hence

$$f(L_i) = c_0 + c_1 L_i + c_2 L_i^2 + \ldots + c_{n-1} L_i^{n-1} \qquad (2.2.1\text{-}7)$$

where we have n equations if we have n distinct eigenvalues which can be used to solve for the c_i's.

A matrix **A** is said to be diagonalizable if **U** exists such that

$$\mathbf{A}' = \mathbf{U}^{-1}\mathbf{A}\mathbf{U} \qquad (2.2.1\text{-}8)$$

where the diagonal terms are the eigenvalues. For A to be diagonalizable, either of the following is sufficient (Hohn, 1957):

a. all of the Li are distinct, or
b. $\mathbf{A}\mathbf{A}^T = \mathbf{A}^T\mathbf{A}$ where $\mathbf{A}^T = \mathbf{A}^*$ (*Hermitian Adjoint*)

A Hermitian matrix is a matrix which is equal to the transpose of its conjugate; that is, $\mathbf{A} = (\mathbf{A}^*)^T$ or $a_{ij} = a_{ji}^*$. As an example consider the matrix A given by

$$\mathbf{A} = \begin{bmatrix} 1 & z \\ z & 1 \end{bmatrix} \qquad (2.2.1\text{-}9)$$

where the eigenvalues are obtained from $|\mathbf{A} - \mathbf{I}\mathbf{L}| = \mathbf{0}$ and are found to be

$$\begin{aligned} L_1 &= 1 + z \\ L_2 &= 1 - z \end{aligned} \qquad (2.2.1\text{-}10)$$

and the eigenvectors are determined from

$$\mathbf{A}\mathbf{U}^{(i)} = L_i \mathbf{U}^{(i)}$$

where for $L = L_1$

$$\begin{bmatrix} -z & z \\ z & -z \end{bmatrix} \begin{bmatrix} u_1^{(1)} \\ u_2^{(1)} \end{bmatrix} = 0$$

We shall assume that $u_1^{(1)} = u_2^{(1)} = k$ where the value of k is selected for convenience. Also, for convenience we normalize $u^{(1)}$ such that

$$\mathbf{u}^{T(1)}\mathbf{u}^{(1)} = 1 \tag{2.2.1-11}$$

Thus,

$$\mathbf{u}^{T(1)}\mathbf{u}^{(1)} = 2k^2 = 1 \tag{2.2.1-12}$$

In the same manner we solve for $\mathbf{u}^{(2)}$. The corresponding normalized eigenvectors are:

$$L = L_1 = 1 + z \qquad \mathbf{u}^{(1)} = 1/2\begin{bmatrix} 1 \\ 1 \end{bmatrix} \tag{2.2.1-13}$$

$$L = L_2 = 1 - z \qquad \mathbf{u}^{(2)} = 1/2\begin{bmatrix} 1 \\ -1 \end{bmatrix} \tag{2.2.1-14}$$

Note that $\mathbf{u}^{T(2)}\mathbf{u}^{(1)} = 0 = \mathbf{u}^{T(1)}\mathbf{u}^{(2)}$. The matrix $\mathbf{U}$ that diagonalizes matrix is

$$\mathbf{U} = \begin{bmatrix} 1/2 & 1/2 \\ 1/2 & -1/2 \end{bmatrix}$$

Note that

$$\mathbf{A}' = \mathbf{U}^{-1}\mathbf{A}\mathbf{U} = \begin{bmatrix} L_1 & 0 \\ 0 & L_2 \end{bmatrix}$$

2.3 THE EXPONENTIAL MATRIX

2.3.1 First-Order System

In this section we would like to discuss the basic concepts involved in replacing a continuous system with a discrete one.

In order to do so, we will use the simplest first-order system to illustrate some of the important properties of these basic concepts. The ideas expressed here will be extended to multi-degree-of-freedom systems with a vector-matrix notation in the next section.

First, consider a homogeneous first-order system

$$\dot{x} + ax = 0$$

with the initial condition $x(0) = x_0$. The analytical solution to this problem is

$$x(t) = x_0 e^{-at}$$

For a specific example, let

$$a = 0.25; \qquad x_0 = 100$$

This leads to

$$x(t) = 100e^{-0.25t}$$

Using a hand calculator, we can compute the exact solution for $t = 1$ as

$$x(1) = 77.88007831$$

We would now like to replace the continuous differential equation with a discrete time-integration method. The class of time-integration method we are interested in using is called a one-step discrete method and is represented by the formula

$$x_{j+1} = mx_j$$

where x_j represents the value at time = jh where h is the timestep and x_{j+1} represents the value at time = (j+1)h. m is a constant which represents the system parameters and the timestep h.

First, let us find the exact value for m which we will denote as m_e. This is found by examining the exact solution and noticing that

$$x(t+h)=100e^{-0.25(t+h)}$$

or

$$x(t+h)=e^{-0.25h}x(t)$$

which gives us the exact value of m as

$$m_e = e^{-0.25h}; \qquad m_e = e^{-ah}$$

For example, let h = 0.2, then m_e = 0.9512294245, and the discrete solution would be obtained from $x_{j+1} = mx_j$ or

$$x_1 = 100$$
$$x_2 = m_e x_1 = 95.12294245 \quad (t=0.2)$$
$$x_3 = m_e x_2 = 90.48374181 \quad (t=0.4)$$
$$x_4 = m_e x_3 = 86.07079765 \quad (t=0.6)$$
$$x_5 = m_e x_4 = 81.87307531 \quad (t=0.8)$$
$$x_6 = m_e x_5 = 77.88007831 \quad (t=1.0)$$

which is the exact value at t = 1.0.

The main reason for introducing the exact m_e is that we now have something to measure against when we introduce all of the various approximations to m_e. The series expansion for e^{-ah} is (let y = ah)

$$m_e = e^{-ah} = e^{-y} = 1 - y + y^2/2 - y^3/6 + \ldots$$

In the vector-matrix case this expansion is known as the *exponential matrix* and is computed in a similar fashion. Computationally, it is only useful for relatively small systems since the exponential matrix becomes fully populated. For this reason, approximations that take advantage of the system matrix sparsity are often used. See Moler and Loan (1978) for a discussion of several ways to compute the exponential matrix.

The approximations that can be used to replace m_e provide an interesting subject, and for the vector-matrix case there are probably 30 to 50 methods used to approximate m_e (Belytschko and Hughes, 1983; Trujillo, 1975). We will only analyze three of the more common methods: the explicit, the fully implicit, and the Crank–Nicolson.

Two of the important characteristics of any method are *stability* and *accuracy*. Stability is simply the requirement that we do not want the solution to blow up. Remember, we will be taking thousands of steps during a problem. In order for the discrete solution not to blow up, it is necessary that the approximate m be less than 1.0. Actually, it is sufficient that the absolute value be less than 1.0. Thus,

$$m = 0.99 \text{ is okay}$$

$$m = 1.01 \text{ is not okay}$$

$$m = -0.99 \text{ is okay}$$

$$m = -1.01 \text{ is not okay}$$

Incidentally, 1000 steps with m = 1.01 will multiply the initial condition by 20959.0, and 1000 steps with m = 0.99 will multiply the initial condition by 4.3E–05.

Accuracy is measured by comparing the approximation to the series expansion of the exact m_e. If the approximation agrees up to the first term $(1 - y)$, it is called a first-order method; if it agrees up to the second term $(1 - y + y^2/2)$, it is called a second-order method, etc. Once an approximation has been selected, the accuracy can be further controlled by varying the timestep h. There is one other criterion that an approximation must meet and that is *consistency*. Consistency means that the original differential equation must be recovered as h approaches zero.

The simplest approximation is the *explicit* method and is obtained by using the first two terms of the expansion

$$m_1 = 1 - y = 1 - ah$$

It is obviously a first-order method and in order for it to be stable it is necessary that $-1 < m_1 < 1$. This requires that h be in the range

$$0 < h < 2/a$$

The value 2/a is the largest timestep that can be taken without m_1 being less than –1. Obviously an m_1 slightly greater than –1 is stable but not accurate. Methods that put limits on the size of the timestep are called *conditionally stable methods*. For our example, $m_1 = 0.95$, which can be compared with the exact $m_e = 0.9512294245$. Carrying out the discrete operations will lead to a value of x_6 of 77.37809375, which compares fairly well with the exact answer of 77.88007831.

The next simplest method is the *fully implicit* and is represented by

$$m_2 = 1/(1 + y) = 1/(1 + ah)$$

Expanding m_2 in a series gives

$$m_2 = 1 - y + y^2 - y^3 + \ldots$$

This shows that this method is also first order because it only agrees up to the first term of the exact method. A partial agreement with the higher-order terms is not sufficient to label a method. In contrast to the explicit method, the stability requirement for this method is met for *any* value of h. It can be seen that for large values of h, m_2 approaches zero. This asymptotic behavior is sometimes desirable because the exact m_e also approaches zero for large values of h. Methods that allow large timesteps are called *unconditionally stable* methods. For our example $m_2 = 0.9523809524$, which can be compared with the exact $m_e = 0.9512294245$. The interested reader can carry out the calculations for x_6.

The third method we will analyze is the *Crank–Nicolson* method (also known as the Padé(2,2) in the literature) and is represented by

$$m_3 = (1-0.5y)/(1+0.5y) = (1-0.5ah)/(1+0.5ah)$$

Expanding m_3 in a series gives

$$m_3 = 1 - y + y^2/2 - y^3/4 + \ldots$$

This method is second order because it agrees exactly up to the second term of the series expansion. The stability requirement for this method is also met for any value of h, hence it is an unconditionally stable method. For large values of h, m_3 approaches –1, but never becomes less than –1. For our example $m_3 = 0.9512195122$. The value of x at $t = 1.0$ using m_3 gives $x_6 = 77.87602064$, which can be compared with the exact 77.88007831. The additional accuracy of the second-order method is clearly demonstrated.

These examples should help to emphasize the differences between stability and accuracy. Stability is a must — a solution cannot be obtained without it. But once a stable solution is obtained, one must still make sure that it is accurate (usually by varying the timestep h). For a more complete discussion of stability, consistency, and the convergence of the difference approximation to the solution of the differential equation, see Isaacson and Keller (1966).

These analyses of stability and accuracy can be extended to multi-degree-of-freedom systems, converting the system to decoupled first-order systems. This gets us into the area of eigenvalues and eigenvectors, which will be discussed in a later chapter.

2.3.2 Multi-Degree-of-Freedom System

Now let's consider a system of first-order equations given in the form

$$\begin{aligned}
dx_1/dt &= a_{11}(t)x_1 + a_{12}(t)x_2 + \ldots + a_{1n}(t)x_n + f_1(t) \\
dx_2/dt &= a_{21}(t)x_1 + a_{22}(t)x_2 + \ldots + a_{2n}(t)x_n + f_2(t) \\
&\cdots\cdots\cdots\cdots\cdots\cdots\cdots\cdots\cdots\cdots \\
dx_n/dt &= a_{n1}(t)x_1 + a_{n2}(t)x_2 + \ldots + a_{nn}(t)x_n + f_n(t)
\end{aligned} \tag{2.3.2-1}$$

these can be written in matrix form as

$$dx/dt = \mathbf{A}(t)\mathbf{x} + \mathbf{f}(t) \tag{2.3.2-2}$$

where

$$\mathbf{A} = (a_{ij})$$

and

$$dx/dt = \begin{bmatrix} dx_1/dt \\ dx_2/dt \\ \vdots \\ dx_n/dt \end{bmatrix} \tag{2.3.2-3}$$

For an initial value problem this requires us to find a solution $\mathbf{x} = \mathbf{x}(t)$ which satisfies an initial condition $\mathbf{x}(t_0) = \mathbf{c}$, where t_0 is some initial point and

$$\mathbf{c} = \begin{bmatrix} c_1 \\ c_2 \\ \vdots \\ c_n \end{bmatrix} \tag{2.3.2-4}$$

Essentially, any system can be put into matrix form. For example, consider the two linear equations

$$dy/dt = ty + 6z + \cos(t) \tag{2.3.2-5a}$$

$$dz/dt = t^3 z - t^2 y + t - 1 \tag{2.3.2-5b}$$

This system of equations is equivalent to the matrix equation

$$\begin{bmatrix} dy/dt \\ dz/dt \end{bmatrix} = \begin{bmatrix} t & 6 \\ -t^2 & t^3 \end{bmatrix} \begin{bmatrix} y \\ z \end{bmatrix} + \begin{bmatrix} \cos(t) \\ t-1 \end{bmatrix} \tag{2.3.2-6}$$

In general, when a system of differential equations is reduced to matrix form, Equation 2.3.2-2, the matrix $\mathbf{A}(t)$ will depend explicitly on the variable t. However, for certain systems, $\mathbf{A}(t)$ will not vary with t, and hence every element of $\mathbf{A}$ is constant. The system is said to have constant coefficients. Let's consider $\mathbf{A}$ to be constant and seek a solution to the initial value problem given by

$$d\mathbf{x}/dt = \mathbf{A}\mathbf{x} + \mathbf{f}(t), \qquad \mathbf{x}(0) = \mathbf{c} \tag{2.3.2-7}$$

Rewriting Equation 2.3.2-7 as $(d\mathbf{x}/dt - \mathbf{A}\mathbf{x}) = \mathbf{f}(t)$ and premultiplying each side by $e^{-\mathbf{A}t}$ yields

$$e^{-\mathbf{A}t}\left(d\mathbf{x}/dt - \mathbf{A}\mathbf{x}\right) = e^{-\mathbf{A}t}\mathbf{f}(t) \tag{2.3.2-8}$$

We note that

$$d/dt\left(e^{-\mathbf{A}t}\mathbf{x}\right) = e^{-\mathbf{A}t}\, d\mathbf{x}/dt - e^{-\mathbf{A}t}\mathbf{A}\mathbf{x}$$
$$= e^{-\mathbf{A}t}\left(d\mathbf{x}/dt - \mathbf{A}\mathbf{x}\right)$$

Therefore, Equation 2.3.2-8 becomes

$$d/dt\left(e^{-\mathbf{A}t}\mathbf{x}\right) = e^{-\mathbf{A}t}\mathbf{f}(t)$$

and, upon integrating between 0 and t, yields

$$\mathbf{x}(t) = e^{\mathbf{A}t}\mathbf{c} + \int_0^t e^{\mathbf{A}(t-s)}\mathbf{f}(s)\, ds \tag{2.3.2-9}$$

Here $e^{\mathbf{A}t}$ is the exponential matrix defined by

$$e^{\mathbf{A}t} = \mathbf{I} + \mathbf{A}t + (\mathbf{A}t)^2/2! + (\mathbf{A}t)^3/3! + \ldots + (\mathbf{A}t)^n/n! \tag{2.3.2-10}$$

Since the scalar function e^x and the matrix function $e^{\mathbf{A}}$ are defined similarly, we note that they possess similar properties. Some of these are defined as follows

$$\left(e^{A}\right)^{-1} = e^{-A} \tag{2.3.2-11a}$$

$$e^{0} = \mathbf{I} \tag{2.3.2-11b}$$

$$\left(e^{A}\right)^{T} = e^{A^{T}} \tag{2.3.2-11c}$$

$$e^{A}e^{B} = e^{A+B} \tag{2.3.2-11d}$$

To solve Equation 2.3.2-9 on the computer, it is convenient to replace it with an equivalent discrete form. Hence, in Equation 2.3.2-9 the values at time t + h can be related to those at time t by

$$\mathbf{x}(t+h) = e^{A(t+h)}\mathbf{c} + \int_{0}^{t+h} e^{A(t+h-s)}\mathbf{f}(s)\,ds \tag{2.3.2-12}$$

From Equation 2.3.2-9 we have

$$e^{A(t+h)}\mathbf{c} = e^{Ah}\mathbf{x}(t) - \int_{0}^{t} e^{A(t+h-s)}\mathbf{f}(s)\,ds \tag{2.3.2-13}$$

Substituting Equation 2.3.2-13 into Equation 2.3.2-12 yields

$$\mathbf{x}(t+h) = e^{Ah}\mathbf{x}(t) + e^{Ah}e^{At}\int_{t}^{t+h} e^{-As}\mathbf{f}(s)\,ds \tag{2.3.2-14}$$

No approximations have been introduced in Equation 2.3.2-14. We will look at approximations to the forcing function and approximations to the exponential matrix in Section 2.4.

2.3.3 Reduction of Higher-Order Systems

There are many differential equations encountered in engineering and science whose order is higher than the first order. These can be solved in the same manner as that presented above by first reducing the equations to a system of first-order ones. Let's consider first, as an example, an equation of third order. Consider:

$$d^3x/dt^3 + 4d^2x/dt^2 + 5dx/dt = t \tag{2.3.3-1}$$

with initial conditions

$$x(0) = c_1, \quad dx(0)/dt = c_2, \quad d^2x(0)/dt^2 = c_3 \tag{2.3.3-2}$$

letting

$$dx/dt = z_1$$

$$dz_1/dt = z_2$$

Equation 2.3.3-1 can be written as

$$\begin{aligned} dx/dt &= z_1 \\ dz_1/dt &= z_2 \\ dz_2/dt &= t - 4z_2 - 5z_1 \end{aligned} \tag{2.3.3-3}$$

with the initial conditions given by

$$x(0) = c_1, \quad z_1(0) = c_2, \quad z_2(0) = c_3 \tag{2.3.3-4}$$

In matrix form Equation 2.3.3-1 can be written

$$dx/dt = \mathbf{Ax} + \mathbf{f}(t)$$

where

$$(\mathbf{x})^T = (x \quad z_1 \quad z_2) \qquad \mathbf{f}(t) = (0 \quad 0 \quad t)$$

$$\mathbf{A} = \begin{bmatrix} 0 & 1 & 0 \\ 0 & 0 & 1 \\ 0 & -5 & -4 \end{bmatrix} \tag{2.3.3-5}$$

We can apply the same technique for a system of equations. Consider the set of equations given by

$$\begin{aligned}
& d^3x_1/dt^3 - d^2x_1/dt^2 - dx_1/dt + x_1 \\
& \quad - d^2x_2/dt^2 + 3dx_2/dt - 2x_2 + dx_3/dt - x_3 = f_1 \\
& 3d^2x_1/dt^2 - 6dx_1/dt + 3x_1 - d^2x_2/dt^2 \\
& \quad + 4dx_2/dt - 3x_2 + 2dx_3/dt - 2x_3 = f_2 \\
& d^2x_1/dt^2 - 2dx_1/dt + x_1 + dx_2/dt - x_2 + dx_3/dt - x_3 = f_3
\end{aligned} \tag{2.3.3-6}$$

with initial conditions

$$x_1(0) = c_1, \quad dx_1(0)/dt = c_2, \quad d^2x_1(0)/dt^2 = c_3$$

$$x_2(0) = c_4, \quad dx_2(0)/dt = c_5, \quad x_3(0) = c_6$$

Letting

$$z_1 = x_1, \quad z_2 = dx_1/dt, \quad z_3 = d^2x_1/dt^2$$

$$z_4 = x_2, \quad z_5 = dx_2/dt, \quad z_6 = x_3$$

then Equation 2.3.3-6 in matrix form becomes

$$dx/dt = \mathbf{Ax} + \mathbf{f}(t)$$

where

$$(\mathbf{x})^T = (z_1 \quad z_2 \quad z_3 \quad z_4 \quad z_5 \quad z_6)$$

$$\mathbf{A} = \begin{bmatrix} 0 & 1 & 0 & 0 & 0 & 0 \\ 0 & 0 & 1 & 0 & 0 & 0 \\ 1 & -3 & 3 & 0 & 0 & 0 \\ 0 & 0 & 0 & 0 & 1 & 0 \\ 1 & -2 & 1 & -1 & 2 & 0 \\ -1 & 2 & -1 & 1 & -1 & 1 \end{bmatrix}$$

$$(f)^T = \begin{pmatrix} 0 & 0 & (f_1 + f_2 + f_3) & 0 & (f_2 - 2f_3) & f_3 \end{pmatrix}$$

and

$$z_1(0) = c_1, \quad z_2(0) = c_2, \quad z_3(0) = c_3$$
$$z_4(0) = c_4, \quad z_5(0) = c_5, \quad z_6(0) = c_6$$

2.3.4 Numerical Example

As a numerical example, let's consider a homogeneous second-order system

$$m\ddot{x} + c\dot{x} + kx = 0 \qquad (2.3.4\text{-}1)$$

with the initial conditions $x(0) = x_0$, $\dot{x}(0) = 0$. Letting

$$2\zeta\omega = c/m \quad \omega^2 = k/m$$

we have

$$\ddot{x} + 2\zeta\omega\dot{x} + \omega^2 = 0 \qquad (2.3.4\text{-}2)$$

The analytical solution is given as

$$x(t) = e^{-\zeta\omega t}\left(A\cos\omega_d t + B\sin\omega_d t\right) \qquad (2.3.4\text{-}3)$$

where for $\zeta < 1$

$$\omega_d = \omega\sqrt{1-\zeta^2}, \quad A = x_0, \quad B = \zeta x_0/\omega_d$$

and Equation 2.3.4-3 becomes

$$x(t) = e^{-\zeta\omega t}\left(x_0\cos\omega_d t + \zeta x_0/\omega_d \sin\omega_d t\right) \qquad (2.3.4\text{-}4)$$

Letting m = 0.025 Lb-sec^2/in, k = 30.625 Lb/in, c = 0.1 Lb-sec/in and x_0 = 10.0 in. gives

$$\zeta\omega = 2 \text{ rad/sec}, \quad \omega = 35 \text{ rad/sec}, \quad \omega_d = 34.943 \text{ rad/sec}$$

and hence Equation 2.3.4-3 can be written

$$x(t) = e^{-2t}\left[10\cos(34.943t) + 0.572\sin(34.943t)\right] \tag{2.3.4-5}$$

We can also obtain the velocity by differentiating Equation 2.3.4-4, thus

$$\dot{x}(t) = -x_0\left(\omega_d + \zeta^2/\omega_d\right)e^{-\zeta\omega t}\sin\left(\omega_d t\right) \tag{2.3.4-6}$$

or

$$\dot{x}(t) = -10(34.943 + 4/34.943)e^{-2t}\sin(34.943t) \tag{2.3.4-7}$$

Using a calculator, we can compute the exact solution for the displacement and velocity at t = 1 as

$$x(1) = -1.2831 \text{ in} \quad \text{and} \quad \dot{x}(1) = 17.839 \text{ in/sec}$$

Using the solution scheme developed above we can apply it to our second-order equation and compare the exact solution obtained. Our second-order equation can be written as two first-order equations in the form

$$\begin{bmatrix} \dot{x} \\ \dot{v} \end{bmatrix} = \begin{bmatrix} 0 & 1 \\ -\omega^2 & -2\zeta\omega \end{bmatrix} \begin{bmatrix} x \\ v \end{bmatrix}$$

which is of the form

$$\mathbf{x} = \mathbf{A}\mathbf{x}$$

where

$$\mathbf{A} = \begin{bmatrix} 0 & 1 \\ -\omega^2 & -2\zeta\omega \end{bmatrix}$$

This equation can be put into discrete form by using Equation 2.3.2-12 with f(s) = 0. Thus, we have

$$\mathbf{x}_{i+1} = e^{\mathbf{A}h}\mathbf{x}_i$$

Let's let h = 0.1; then $\mathbf{M} = e^{\mathbf{A}h}$ is given as

$$\mathbf{M} = \begin{bmatrix} -0.784523 & -0.008093 \\ 9.9144650 & -0.752149 \end{bmatrix}$$

and the discrete solution becomes

$$\mathbf{x}_{i+1} = \mathbf{M}\mathbf{x}_i$$

Carrying out the operations gives the following values:

$$
\begin{array}{lll}
t = 0.0 & x_1 = 10.0 & v_1 = 0.0 \\
t = 0.1 & x_2 = -7.845 & v_2 = 99.145 \\
t = 0.2 & x_3 = 5.352 & v_3 = -152.353 \\
t = 0.3 & x_4 = -2.966 & v_4 = 167.658 \\
t = 0.4 & x_5 = 0.970 & v_5 = -155.509 \\
t = 0.5 & x_6 = 0.498 & v_6 = 126.583 \\
t = 0.6 & x_7 = -1.415 & v_7 = -90.275 \\
t = 0.7 & x_8 = 1.841 & v_8 = 53.872 \\
t = 0.8 & x_9 = -1.880 & v_9 = -22.271 \\
t = 0.9 & x_{10} = 1.655 & v_{10} = -1.889 \\
t = 1.0 & x_{11} = -1.283 & v_{11} = 17.831
\end{array}
$$

which are very close to the exact values at t=1.0.

2.4 APPROXIMATIONS TO THE EXPONENTIAL MATRIX

2.4.1 Padé Approximations

The solution to the set of first-order equations developed in Section 2.3.2 in discrete form was given by

$$\mathbf{x}(t+h)=e^{Ah}\mathbf{x}(t)+e^{Ah}e^{At}\int_{t}^{t+h}e^{-As}\mathbf{f}(s)\,ds \qquad (2.4.1\text{-}1)$$

where no approximations were introduced. It is seen that approximations can be made to both the exponential matrix e^{Ah} and the forcing term f(s). It is important to keep these approximations separate since the nature of the forcing function is usually known, while on the other hand, knowledge of the system time constants is usually lacking.

In the approximations to the forcing function we note that in Equation 2.4.1-1 the integrand consists of two terms — the exponential matrix and the forcing function. We can approximate the entire integrand; or better yet we can approximate only the forcing function and then complete the integration involving the exponential matrix. We also have a choice as to the type of approximation to use for **f**(s). A frequently used approximation, especially useful in discrete control processes (Lapidus and Luus, 1967) is to consider **f**(s) as being constant in the interval (t,t+h). Considering that f(s) is approximated by the value of **f** at the beginning of the interval we have

$$\mathbf{f}(s)\approx\mathbf{f}_j \quad \text{in } t\le s\le t+h \qquad (2.4.1\text{-}2)$$

which results in the following equation

$$\mathbf{x}_{j+1}=e^{Ah}\mathbf{x}_j+\left(\mathbf{A}^{-1}e^{Ah}-\mathbf{A}^{-1}\right)\mathbf{f}_j \qquad (2.4.1\text{-}3)$$

where $\mathbf{x}_j$ denotes the solution at time equal to jh. Equation 2.4.1-3 can be written as

$$\mathbf{x}_{j+1}=\mathbf{M}\mathbf{x}_j+\mathbf{P}\mathbf{f}_j \qquad (2.4.1\text{-}4)$$

where

$$\mathbf{M} = e^{\mathbf{A}h}, \quad \mathbf{P} = (\mathbf{M} - \mathbf{I})\mathbf{A}^{-1} \tag{2.4.1-5}$$

Another useful formula can be obtained if **f**(t) is approximated by a set of piecewise linear segments. In each interval (t,t+h) we represent **f**(t) by

$$\mathbf{f}(s) \approx \mathbf{f}_j + (s - t)/h(\mathbf{f}_{j+1} - \mathbf{f}_j) \quad \text{in } t \leq s \leq t + h$$

Substituting this expression into Equation 2.4.1-1 and completing the integration involving the exponential matrix yields

$$\begin{aligned} \mathbf{x}_{j+1} &= e^{\mathbf{A}h}\mathbf{x}_j + (e^{\mathbf{A}h} - \mathbf{I})\mathbf{A}^{-1}\mathbf{f}_{j+1} \\ &+ \left[e^{\mathbf{A}h} - (e^{\mathbf{A}h} - \mathbf{I})\mathbf{A}^{-1}\right]\mathbf{A}^{-1}(\mathbf{f}_j - \mathbf{f}_{j+1})/h \end{aligned} \tag{2.4.1-6}$$

Letting

$$\mathbf{D} = \left[e^{\mathbf{A}h} - (e^{\mathbf{A}h} - \mathbf{I})\mathbf{A}^{-1}\right]\mathbf{A}^{-1}/h \tag{2.4.1-7}$$

then Equation 2.4.1-6 becomes

$$\mathbf{x}_{j+1} = \mathbf{M}\mathbf{x}_j + \mathbf{P}\mathbf{f}_{j+1} + \mathbf{D}(\mathbf{f}_j - \mathbf{f}_{j+1}) \tag{2.4.1-8}$$

Some of the advantages of Equation 2.4.1-8 are that the only approximation errors (not including truncation and roundoff) are due to the piecewise linear approximation to **f**(t). It is unconditionally stable with respect to the step-size h, and we can obtain exact answers for the homogeneous or for the nonhomogeneous case where **f**(t) is truly a piecewise linear function. However, Equation 2.4.1-8 has some disadvantages. These include the following. Each step requires n^2 multiplications, changes in step-size h or the system matrix **A** require considerable computational effort, and the analysis requires storage of three full $n \times n$ matrices.

A major part of the computational effort in the above formulation is in the evaluation of the infinite series for e^{Ah} as given by Equation 2.3.2-10. The selection of a step-size h_0 to actually use in numerically evaluating the infinite series depends on the number of digits available in the computer and the nature of the matrix **A**. A discussion of how to select an appropriate step-size to use in the evaluation of the series can be found in (Moler and Loan, 1978). Once the magnitude of h_0 has been determined, the exponential matrix can be calculated for any h, by choosing m so that

$$h_0 = h/2^m$$

The matrix e^{Ah_0} is then calculated. To obtain e^{Ah}, first square the matrix e^{Ah_0} and then square the resulting matrix and continue the process until the desired result is reached. The process is

$$e^{Ah_0} e^{Ah_0} = e^{2Ah_0}$$

$$e^{2Ah_0} e^{2Ah_0} = e^{4Ah_0}, \text{ etc.}$$

In this manner, large values of h can be reached in relatively few steps.

Approximations to the exponential matrix are called the *Padé matrix approximations* (Varga, 1962; Trujillo, 1975), and several have been listed in Table 2.1. Also shown, in brackets, are the lowest order terms that deviate from the correct series. The application of these approximations to the inhomogeneous case are listed in Table 2.2.

The common forward and backward differencing schemes are represented by equations F1 and B1. Equation E1 is more commonly known as the Crank–Nicolson method, which we shall discuss in the next section. The forward or explicit approximations are the only ones that do not require solving a set of simultaneous equations. However, they have the limitation of being stable only for certain magnitudes of step-size, whereas the implicit methods are unconditionally stable. The backward or implicit approximations offer no significant computational advantages over the even combinations. They both involve a similar inverse, and as Table 2.1 shows, the even combinations

TABLE 2.1

Padé Exponential Matrix Approximations

$e^{Ah} = \mathbf{I} + \mathbf{A}h + (\mathbf{A}h)^2/2! + \ldots + (\mathbf{A}h)^n/n! + \ldots$

Forward (explicit)

F1. $\mathbf{I} + \mathbf{A}h$

F2. $\mathbf{I} + \mathbf{A}h + (\mathbf{A}h)^2/2$

Backward (implicit)

B1. $(\mathbf{I} - \mathbf{A}h)^{-1} = \mathbf{I} + \mathbf{A}h + [(\mathbf{A}h)^2] + \ldots$

B2. $(\mathbf{I} - \mathbf{A}h + (\mathbf{A}h)^2/2)^{-1} = \mathbf{I} + \mathbf{A}h + (\mathbf{A}h)^2/2 - [(\mathbf{A}h)^4/4] + \ldots$

Uneven combination

U1. $(\mathbf{I} - \mathbf{A}h/3)^{-1}(\mathbf{I} + 2/3\mathbf{A}h + (\mathbf{A}h)^2/6) = \mathbf{I} + \mathbf{A}h$
$+ (\mathbf{A}h)^2/2 + (\mathbf{A}h)^3/3! + [(\mathbf{A}h)^4/18] + \ldots$

Even combination

E1. $(\mathbf{I} - \mathbf{A}h/2)^{-1}(\mathbf{I} + \mathbf{A}h/2) = \mathbf{I} + \mathbf{A}h + (\mathbf{A}h)^2/2 + [(\mathbf{A}h)^3/4] + \ldots$

E2. $(\mathbf{I} - \mathbf{A}h/2 + (\mathbf{A}h)^2/12)^{-1}(\mathbf{I} + \mathbf{A}h/2 + (\mathbf{A}h)^2/12) = \mathbf{I} + \mathbf{A}h + \ldots$
$+ [(\mathbf{A}h)^5/144] + \ldots$

E3. $(\mathbf{I} - \mathbf{A}h/2 + (\mathbf{A}h)^2/10 - (\mathbf{A}h)^3/120)^{-1}(\mathbf{I} + \mathbf{A}h/2 + (\mathbf{A}h)^2/10$
$+ (\mathbf{A}h)^3/120) = \mathbf{I} + \mathbf{A}h + \ldots + [(\mathbf{A}h)^7/4{,}800] + \ldots$

E4. $(\mathbf{I} - \mathbf{A}h/2 + 3/28(\mathbf{A}h)^2 - (\mathbf{A}h)^3/84 + (\mathbf{A}h)^4/1{,}680)^{-1}(\mathbf{I} + \mathbf{A}h/2 + 3/28(\mathbf{A}h)^2$
$+ (\mathbf{A}h)^3/84 + (\mathbf{A}h)^4/1{,}680) = \mathbf{I} + \mathbf{A}h + \ldots + [1.969 \times 10^{-5}/1.69344(\mathbf{A}h)^9]$

TABLE 2.2

Approximations to Inhomogeneous Case (Equation 2.4.1-6)

Forward (explicit)

F1. $\mathbf{x}_{n+1} = (\mathbf{I} + \mathbf{A}h)\mathbf{x}_n + h\mathbf{f}_n$

F2. $\mathbf{x}_{n+1} = (\mathbf{I} + \mathbf{A}h + (\mathbf{A}h)^2/2)\mathbf{x}_n + (\mathbf{f}_{n+1} + \mathbf{f}_n)h/2 + \mathbf{A}\mathbf{f}_n h^2/2$

Backward (implicit)

B1. $(\mathbf{I} - \mathbf{A}h)\mathbf{x}_{n+1} = \mathbf{x}_n + h\mathbf{f}_{n+1}$

B2. $(\mathbf{I} - \mathbf{A}h + (\mathbf{A}h)^2/2)\mathbf{x}_{n+1} = \mathbf{x}_n + (\mathbf{f}_{n+1} + \mathbf{f}_n)h/2 - \mathbf{A}\mathbf{f}_{n+1}h^2/2$

Uneven combination

U1. $(\mathbf{I} - \mathbf{A}h/3)\mathbf{x}_{n+1} = (\mathbf{I} + 2/3\mathbf{A}h + (\mathbf{A}h)^2/6)\mathbf{x}_n + (\mathbf{f}_{n+1} + \mathbf{f}_n)h/2 + \mathbf{A}\mathbf{f}_n h^2/6$

Even combinations

E1. $(\mathbf{I} - \mathbf{A}h/2)\mathbf{x}_{n+1} = (\mathbf{I} + \mathbf{A}h/2)\mathbf{x}_n + (\mathbf{f}_{n+1} + \mathbf{f}_n)h/2$

E2. $(\mathbf{I} - \mathbf{A}h/2 + (\mathbf{A}h)^2/12)\mathbf{x}_{n+1} = (\mathbf{I} + \mathbf{A}h/2 + (\mathbf{A}h)^2/12)\mathbf{x}_n$
$+ (\mathbf{f}_{n+1} + \mathbf{f}_n)h/2 + \mathbf{A}(\mathbf{f}_n - \mathbf{f}_{n+1})h^2/12$

E3. $(\mathbf{I} - \mathbf{A}h/2 + (\mathbf{A}h)^2/10 - (\mathbf{A}h)^3/120)\mathbf{x}_{n+1} = (\mathbf{I} + \mathbf{A}h/2 + (\mathbf{A}h)^2/10$
$+ (\mathbf{A}h)^3/120)\mathbf{x}_n + (\mathbf{f}_{n+1} + \mathbf{f}_n)h/2 + \mathbf{A}(\mathbf{f}_n - \mathbf{f}_{n+1})11h^2/120$
$+ \mathbf{A}^2(\mathbf{f}_n + \mathbf{f}_{n+1})h^3/120$

represent a higher-order approximation. The implicit approximations do, however, tend to behave better for very large timesteps, since they will approach a quasi-steady-state value for large values of h.

The Crank–Nicolson method (Equation E1 in Table 2.1) is an implicit technique that uses the average value of the derivative to move the solution ahead in time. It is noted that for a slight decrease in accuracy one can use larger timesteps with this method to reduce the number of computations. However, as one increases the timestep, oscillations in the Crank–Nicolson method may become more severe, but they will never become unstable as would happen using an explicit procedure such as Equations F1 and F2 in Table 2.1.

To demonstrate the derivation of the Crank–Nicolson formulas, consider Equation 2.4.1-3, which was given as

$$\mathbf{x}_{j+1} = e^{\mathbf{A}h}\mathbf{x}_j + \left(\mathbf{A}^{-1}e^{\mathbf{A}h} - \mathbf{A}^{-1}\right)\mathbf{f}_j \tag{2.4.1-3}$$

Using Equation E1 from Table 2.1 we have

$$e^{\mathbf{A}h} \approx \left(\mathbf{I} - \mathbf{A}h/2\right)^{-1}\left(\mathbf{I} + \mathbf{A}h/2\right)$$

Therefore,

$$\begin{aligned} \mathbf{x}_{j+1} &= \left(\mathbf{I} - \mathbf{A}h/2\right)^{-1}\left(\mathbf{I} + \mathbf{A}h/2\right)\mathbf{x}_j \\ &+ \left(\left[\mathbf{I} - \mathbf{A}h/2\right]^{-1}\left[\mathbf{I} + \mathbf{A}h/2\right] - \mathbf{I}\right)\mathbf{A}^{-1}\mathbf{f}_j \end{aligned} \tag{2.4.1-9}$$

which can be written

$$\begin{aligned} \left(\mathbf{I} - \mathbf{A}h/2\right)\mathbf{x}_{j+1} &= \left(\mathbf{I} + \mathbf{A}h/2\right)\mathbf{x}_j \\ &+ \left(\left[\mathbf{I} + \mathbf{A}h/2\right] - \left[\mathbf{I} - \mathbf{A}h/2\right]\right)\mathbf{A}^{-1}\mathbf{f}_j \end{aligned} \tag{2.4.1-10}$$

or finally

$$\left(\mathbf{I} - \mathbf{A}h/2\right)\mathbf{x}_{j+1} = \left(\mathbf{I} + \mathbf{A}h/2\right)\mathbf{x}_j + h\mathbf{f}_j \tag{2.4.1-11}$$

Applying the same technique to Equation 2.4.1-6 leads to the following expression

$$(\mathbf{I} - \mathbf{A}h/2)\mathbf{x}_{j+1} = (\mathbf{I} + \mathbf{A}h/2)\mathbf{x}_j + (\mathbf{f}_{j+1} + \mathbf{f}_j)h/2 \qquad (2.4.1\text{-}12)$$

Exercise

Show that it is not necessary for $\mathbf{A}^{-1}$ to exist in order to compute the matrix **P** in Equation 2.4.1-5.

Exercise

Outline a procedure to calculate **P** from a series using the relationship $h_0 = h/2^m$ where h_0 will be used in the series and h is the desired timestep.

2.4.2 Structural Dynamics

A common form of matrix linear differential equation found in the area of structural dynamics is the second-order equation

$$\mathbf{M}\ddot{\mathbf{y}} + \mathbf{C}\dot{\mathbf{y}} + \mathbf{K}\mathbf{y} = \mathbf{g}(t) \qquad (2.4.2\text{-}1)$$

where **M**, **C**, and **K** are the mass, damping, and stiffness matrices, respectively. **y** is the displacement, and **g** is the forcing term. There have been many studies (Chan, 1962; Bathe, 1973) of approximate methods to integrate Equation 2.4.2-1, especially for large systems of equations with banded matrices. Equation 2.4.2-1 can be represented as a system of first-order equations using the method given in Section 2.3.2 with the following definitions:

$$\mathbf{x} = \begin{bmatrix} \mathbf{y} \\ \mathbf{v} \end{bmatrix} \quad \mathbf{A} = \begin{bmatrix} \mathbf{0} & \mathbf{I} \\ -\mathbf{M}^{-1}\mathbf{K} & -\mathbf{M}^{-1}\mathbf{C} \end{bmatrix} \quad \text{and} \quad \mathbf{f} = \begin{bmatrix} \mathbf{0} \\ \mathbf{g} \end{bmatrix} \qquad (2.4.2\text{-}2)$$

where **v** represents the velocity vector $\dot{\mathbf{y}}$. Equation 2.4.2-1 becomes

$$\begin{bmatrix} \dot{\mathbf{y}} \\ \dot{\mathbf{v}} \end{bmatrix} = \begin{bmatrix} \mathbf{0} & \mathbf{I} \\ -\mathbf{M}^{-1}\mathbf{K} & -\mathbf{M}^{-1}\mathbf{C} \end{bmatrix} \begin{bmatrix} \mathbf{y} \\ \mathbf{v} \end{bmatrix} + \begin{bmatrix} \mathbf{0} \\ \mathbf{g} \end{bmatrix} \qquad (2.4.2\text{-}3)$$

The approximations given in Table 2.2 can now be applied directly. The matrix **A** is defined in Equation 2.4.2-2, and the vectors $\mathbf{x}_n$ and $\mathbf{f}_n$ are

$$\mathbf{x}_n = \begin{bmatrix} \mathbf{y}_n \\ \mathbf{v}_n \end{bmatrix} \quad \mathbf{f}_n = \begin{bmatrix} \mathbf{0} \\ \mathbf{g}_n \end{bmatrix}$$

With the above definitions and a little algebra, one can arrive at the formulas in Table 2.3. In arriving at several of the formulas in Table 2.3, a group of terms occurred which represented the acceleration defined as

$$\mathbf{M}\mathbf{a}_n = \mathbf{g}_n - \mathbf{C}\mathbf{v}_n - \mathbf{K}\mathbf{y}_n \qquad (2.4.2\text{-}4)$$

In these cases, the use of the acceleration as a state variable resulted in a considerable simplification of the formulas. One of the reasons for using the special form of the dynamical equations is to reduce the size of the system matrix so that the state vectors $\mathbf{y}_{n+1}$ and $\mathbf{v}_{n+1}$ can be solved for independently. As Table 2.3 shows, this was accomplished in all of the formulas except B2 and E2. For these, two system matrices must be stored to advance the solution. This is not unexpected since B2 and E2 are higher-order approximations.

The formula E1 in Table 2.3 is also known as *Newmark's generalized acceleration method* with $\theta = 1/2$, and $\alpha = 1/4$ (Hughes, 1987). Although we cannot completely reduce Equations E2, we can combine them to obtain the following

$$\begin{aligned} \left(\mathbf{M} + \mathbf{C}h/2 + \mathbf{K}h^2/6\right)\mathbf{v}_{n+1} &= \left(\mathbf{M} - \mathbf{K}h^2/3\right)\mathbf{v}_n - \mathbf{K}\mathbf{y}_n h/2 \\ &+ \mathbf{M}h/2\mathbf{a}_n + \mathbf{g}_{n+1}h/2 + \left(\mathbf{C}h^2/12 + \mathbf{K}h^3/24\right)\left(\mathbf{a}_{n+1} - \mathbf{a}_n\right) \end{aligned} \qquad (2.4.2\text{-}5)$$

In Equation 2.4.2-5, we have placed the unknown acceleration $\mathbf{a}_{n+1}$ on the right-hand side. Equation 2.4.2-5 is very similar to some well-known formulas. If the term a_{n+1} is dropped along

TABLE 2.3

Application of Approximations to Structural Dynamic Systems

Forward (explicit)

F1. $\mathbf{y}_{n+1} = \mathbf{y}_n + h\mathbf{v}_n$

$\mathbf{M}\mathbf{v}_{n+1} = -\mathbf{K}\mathbf{y}_n h + (\mathbf{M} - \mathbf{C}h)\mathbf{v}_n + \mathbf{g}_n$

F2. $\mathbf{y}_{n+1} = \mathbf{y}_n + h\mathbf{v}_n + \mathbf{a}_n h^2/2$

$\mathbf{M}\mathbf{v}_{n+1} = -\mathbf{K}\mathbf{y}_n h + (\mathbf{M} - \mathbf{C}h - \mathbf{K}h^2/2)\mathbf{v}_n + (\mathbf{g}_{n+1} + \mathbf{g}_n)h/2 - \mathbf{C}\mathbf{a}_n h^2/2$

where

$\mathbf{a}_n = \mathbf{M}^{-1}(\mathbf{g}_n - \mathbf{K}\mathbf{y}_n - \mathbf{C}\mathbf{v}_n)$

Backward (implicit)

B1. $(\mathbf{M} + \mathbf{C}h + \mathbf{K}h^2)\mathbf{v}_{n+1} = \mathbf{M}\mathbf{v}_n - \mathbf{K}\mathbf{y}_n h + h\mathbf{g}_{n+1}\,;\ \mathbf{y}_{n+1} = \mathbf{y}_n + h\mathbf{v}_{n+1}$

B2. $(\mathbf{M} - \mathbf{K}h^2/2)\mathbf{a}_{n+1} + (\mathbf{C} + \mathbf{K}h)\mathbf{v}_{n+1} = \mathbf{M}\mathbf{a}_n + \mathbf{C}\mathbf{v}_n - (\mathbf{g}_{n+1} + \mathbf{g}_n)$

$-(\mathbf{M}h + \mathbf{C}h^2/2)\mathbf{a}_{n+1} + (\mathbf{M} - \mathbf{K}h^2/2)\mathbf{v}_{n+1} = \mathbf{M}\mathbf{v}_n + (\mathbf{g}_n - \mathbf{g}_{n+1})h/2$

$\mathbf{y}_{n+1} = \mathbf{y}_n + h\mathbf{v}_{n+1} - h^2/2\mathbf{a}_{n+1}$

Uneven combination

U1. $\mathbf{y}_{n+1} - h/3\mathbf{v}_{n+1} = \mathbf{y}_n + \mathbf{v}_n 2h/3 + \mathbf{a}_n h^2/6$

$\mathbf{K}h/3\mathbf{y}_{n+1} + (\mathbf{M}+\mathbf{C}h/3)\mathbf{v}_{n+1} = -2h\mathbf{K}/3\mathbf{y}_n + (\mathbf{M} - 2h/3\mathbf{C} - \mathbf{K}h^2/6)\mathbf{v}_n$
$- \mathbf{C}\mathbf{a}_n h^2/6 + (\mathbf{g}_{n+1} + \mathbf{g}_n)h/2$

The above can be combined to eliminate $\mathbf{y}_{n+1}$

$(\mathbf{M} + \mathbf{C}h/3 + \mathbf{K}h^2/9)\mathbf{v}_{n+1} = -\mathbf{K}\mathbf{y}_n + (\mathbf{M} - 2h/3\mathbf{C} - 7/18h^2\mathbf{K})\mathbf{v}_n$
$- (\mathbf{K}h^3/18 + \mathbf{C}h^2/6)\mathbf{a}_n + (\mathbf{g}_{n+1} + \mathbf{g}_n)h/2$

Even combinations

E1. $\mathbf{y}_{n+1} - h/2\mathbf{v}_{n+1} = \mathbf{y}_n + h/2\mathbf{v}_n$

$\mathbf{K}h/2\mathbf{y}_{n+1} + (\mathbf{M} + \mathbf{C}h/2)\mathbf{v}_{n+1} = -\mathbf{K}h/2\mathbf{y}_n + (\mathbf{M} - \mathbf{C}h/2)\mathbf{v}_n + (\mathbf{g}_{n+1} + \mathbf{g}_n)h/2$

The above can be combined to eliminate $\mathbf{y}_{n+1}$

$(\mathbf{M} + \mathbf{C}h/2 + \mathbf{K}h^2/4)\mathbf{v}_{n+1} = -\mathbf{K}\mathbf{y}_n h + (\mathbf{M} - \mathbf{C}h/2 - \mathbf{K}h^2/4)\mathbf{v}_n$
$+ (\mathbf{g}_{n+1} + \mathbf{g}_n)h/2$

E2. $(\mathbf{M} - \mathbf{K}h^2/12)\mathbf{a}_{n+1} + (\mathbf{C} + \mathbf{K}h/2)\mathbf{v}_{n+1} = (\mathbf{M} - \mathbf{K}h^2/12)\mathbf{a}_n + (\mathbf{C} - \mathbf{K}h/2)\mathbf{v}_n$
$+ (\mathbf{g}_{n+1} - \mathbf{g}_n)\ - (\mathbf{M}h/2 + \mathbf{C}h^2/12)\mathbf{a}_{n+1} + (\mathbf{M} - \mathbf{K}h^2/12)\mathbf{v}_{n+1}$
$= (\mathbf{M} - \mathbf{K}h^2/12)\mathbf{v}_n + (\mathbf{M}h/2 - \mathbf{C}h^2/12)\mathbf{a}_n$

$\mathbf{y}_{n+1} = \mathbf{y}_n + h/2(\mathbf{v}_{n+1} + \mathbf{v}_n) + h^2/12(\mathbf{a}_n - \mathbf{a}_{n+1})$

with the $\mathbf{C}h^2/12$ term and part of the $\mathbf{K}h^3/24$ term, we can arrive at the Wilson θ method formula for $\theta = 1$, which is given by

$$\left(\mathbf{M}+\mathbf{C}h/2+\mathbf{K}h^2/6\right)\mathbf{v}_{n+1} = \left(\mathbf{M}-\mathbf{K}h^2/3\right)\mathbf{v}_n - \mathbf{K}\mathbf{y}_n h/2 + \left(\mathbf{M}h/2-\mathbf{K}h^3/12\right)\mathbf{a}_n + \mathbf{g}_{n+1}h/2 \tag{2.4.2-6}$$

Equation 2.4.2-6 is also the same as Newmark's generalized acceleration method with $\theta = 1/2$ and $\alpha = 1/6$. The effect of dropping the terms in Equation 2.4.2-5 will reduce the order of the integration. Equations E2 in Table 2.3 are of the order of h^4 as indicated in Table 2.1.

2.5 EIGENVALUE REDUCTION

Sometimes it is efficient to approximate a large-scale model with a smaller number of degrees of freedom. This can be accomplished using an eigenvalue reduction technique (Clough, 1975). Consider the second-order dynamic system given by

$$\mathbf{M}\ddot{\mathbf{x}} + \mathbf{C}\dot{\mathbf{x}} + \mathbf{K}\mathbf{x} = \mathbf{T}\mathbf{f}(t) \tag{2.5-1}$$

where

$\mathbf{x}$ = (n × 1) generalized displacements
$\mathbf{f(t)}$ = (n_f × 1) forcing function
$\mathbf{T}$ = (n × n_f) location matrix
$\mathbf{M}$ = (n × n) mass matrix
$\mathbf{C}$ = (n × n) damping matrix
$\mathbf{K}$ = (n × n) stiffness matrix

Assume that an eigenvalue analysis has been performed and that each distinct eigenvector ϕ_i and eigenvalue λ_i satisfy the following relationships

$$\mathbf{K}\phi_i = \mathbf{M}\phi_i\lambda_i \tag{2.5-2}$$

$$(\mathbf{K}\phi_i, \phi_i) = 1.0 \tag{2.5-3}$$

$$(\mathbf{K}\phi_i, \phi_i) = \lambda_i^2 \tag{2.5-4}$$

Furthermore, we assume that the following orthogonality conditions are satisfied

$$(\mathbf{K}\phi_i, \phi_j) = 0, \quad (\mathbf{M}\phi_i, \phi_j) = 0, \quad \text{for } i \neq j$$

These modal vectors can be combined into a single matrix

$$\phi = \begin{bmatrix} \phi_1 & \phi_2 & \cdots & \phi_n \end{bmatrix} \tag{2.5-5}$$

In addition, the modal components can be expressed in the following matrix form

$$\phi^T \mathbf{K} \phi = \begin{bmatrix} \lambda_1^2 & & & \\ & \lambda_2^2 & & \\ & & \ddots & \\ & & & \lambda_n^2 \end{bmatrix} = \lambda \tag{2.5-6}$$

$$\phi^T \mathbf{M} \phi = \mathbf{I} \tag{2.5-7}$$

We also assume that the damping matrix satisfies

$$\phi^T \mathbf{C} \phi = b\lambda \tag{2.5-8}$$

The original system can now be transformed using

$$\mathbf{x} = \phi \mathbf{z} \tag{2.5-9}$$

where $\mathbf{z}$ is an $(n \times 1)$ vector. Substituting Equation 2.5-9 into Equation 2.5-1 gives

$$\mathbf{M}\phi\ddot{\mathbf{z}} + \mathbf{C}\phi\dot{\mathbf{z}} + \mathbf{K}\phi\mathbf{z} = \mathbf{Tf} \tag{2.5-10}$$

Premultiplying by ϕ^T results in the decoupled set of equations

$$\ddot{\mathbf{z}} + b\lambda\dot{\mathbf{z}} + \lambda\mathbf{z} = \phi^T \mathbf{Tf} \tag{2.5-11}$$

Equation 2.5-9 can be thought of as a linear combination of modal components with z_i as the weighting factors. That is,

$$\mathbf{x} = z_1\phi_1 + z_2\phi_2 + z_3\phi_3 + \ldots + z_n\phi_n \tag{2.5-12}$$

Now these modal components can be used to reduce the order of the system by using only a few select ones to approximate $\mathbf{x}$.

The question remains as to how many and which of the eigenvectors can adequately represent **x**(t). If the criterion is to include only the lower frequency modes, then the magnitude of natural frequencies can be used to eliminate the higher modes. In some cases, however, it may be possible to eliminate some lower modes if it can be shown that they do not "participate" in the simulation.

The use of a few selected modes to approximate the state variables is represented as

$$\mathrm{x} = \mathbf{Sp} \tag{2.5-13}$$

where S is an $(n \times m)$ matrix contains the selected modes and **p** is an $m \times 1$ vector representing m modes. This gives

$$\ddot{\mathbf{p}} + \mathrm{b}\lambda\dot{\mathbf{p}} + \lambda\mathbf{p} = \mathbf{S}^{\mathrm{T}}\mathbf{Tf} \tag{2.5-14}$$

This system can be converted to a system of first-order equations:

$$\begin{bmatrix} \mathrm{x}_1 \\ \mathrm{x}_2 \end{bmatrix} = \begin{bmatrix} \dot{\mathbf{p}} \\ \mathbf{p} \end{bmatrix} \tag{2.5-15}$$

$$\begin{bmatrix} \dot{\mathrm{x}}_1 \\ \dot{\mathrm{x}}_2 \end{bmatrix} = \begin{bmatrix} -\mathrm{b}\lambda & -\lambda \\ \mathbf{I} & 0 \end{bmatrix} \begin{bmatrix} \mathrm{x}_1 \\ \mathrm{x}_2 \end{bmatrix} = \begin{bmatrix} \mathbf{S}^{\mathrm{T}}\mathbf{Tf} \\ \mathbf{0} \end{bmatrix} \tag{2.5-16}$$

These equations are in a form that can be used for the expressions developed in previous sections.

Chapter 3

THE GENERAL INVERSE PROBLEM

3.1 INTRODUCTION

Engineering problems can be broadly categorized into four general areas:

1. Direct problems
2. Inverse problems
3. Systems identification problems
4. Research problems

For the general multidimensional system

$$\mathbf{x}_{j+1} = \mathbf{M}\mathbf{x}_j + \mathbf{P}\mathbf{g}_j, \quad \mathbf{x}_1 = \mathbf{c} \tag{3.1-1}$$

these areas correspond to the following situations:

1. Direct problem: the system matrices **M** and **P**, the initial conditions **c**, and the external forces $\mathbf{g}_j$'s are known. Find the state variables $\mathbf{x}_j$'s.
2. Inverse problem: **M** and **P** are known together with some measurements related to the $\mathbf{x}_j$'s. Find the $\mathbf{g}_j$'s and **c**.
3. Systems identification problems: **c** and $\mathbf{g}_j$ are known together with some measurements related to the $\mathbf{x}_j$'s. Find **M** and **P** or some components of **M** and **P**.
4. Research problems: some measurements related to the $\mathbf{x}_j$'s have been taken. Find **M**, **P**, **c**, and the $\mathbf{g}_j$'s.

The direct problem is by far the more common and is the subject of most engineering textbooks. It should be pointed out that it does not involve measurements. Thus, models are constructed assuming the existence of all orders of derivatives and smoothness. However, since measurements always have some degree of noise and uncertainty, it becomes a major task to join models and measurements.

The systems identification problem (also known as the parameter estimation problem) is the subject of much research. The more successful approaches are those in which the general form of **M** and **P** are known and only a few parameters are unknown. See Chapter 8 for an introduction to some techniques.

The research problem is clearly the most difficult since not only is the system unknown, but so also are the external forces. These problems are usually solved by painstaking experimentation or extraordinary insight. A classic example is Isaac Newton's contribution to celestial mechanics.

In this book we will be concerned mainly with inverse problems. With the power of the computer, it is now possible to obtain practical solutions, and it is not too surprising that it takes a combination of four rather abstract mathematical concepts to provide even a start toward solving this class of problem. These are:

1. Least squares criteria with regularization (sometimes referred to as Tikhonov's method)
2. Dynamic programming to provide a solution
3. Generalized cross validation to optimally estimate the smoothing parameter
4. Chandrasekhar equations to allow the solution of large systems.

The first two concepts are presented in Chapter 1. The last two will be discussed in this chapter.

3.2 GENERALIZED CROSS VALIDATION

This method represents a very simple and practical idea. While many methods exist for estimating parameters in the presence of noise, most require *a priori* knowledge of the noise statistics. In most situations this knowledge is lacking. The method of cross validation will allow us to optimally estimate parameters by using the data and only the data itself. For a more statistical explanation, see Golub (1979), Wahba (1990), and Woltring (1985).

Cross validation is based on the idea of removing one of the data points from the set and then solving the inverse problem using all of the remaining data points. This will give one a solution (estimate) for all of the state variables including the one corresponding to the deleted data point. Since this data point was not included, it can be considered to be an independent observation. One can now compare the estimated state variable against the deleted point and in this way obtain a measure (error) of the estimate. Now repeat this process for another data point, then another, etc. The sum of the squares of all these predicted errors is then a measure of how well the proposed model performs. In the case of the inverse problem, the smoothing parameter is the only remaining model parameter. The value of the smoothing parameter that *minimizes* the predicted errors is the optimal one.

Although the idea is simple, it would not be very practical if one had to proceed in a brute force manner and solve each problem separately, deleting one data point at a time. Fortunately, it will be shown that the computation of the predicted errors is relatively simple, and more important, it will be shown that only a few extra computations are involved in the dynamic programming formulas. The origin of the ordinary cross validation estimate is considered important and will be derived in the next section. It requires solving a set of three related optimization problems.

First, consider the static inverse problem of estimating the state $\mathbf{x}$ from a set of measurements $\mathbf{d}$. The state $\mathbf{x}$ is related to a control vector $\mathbf{g}$ by

$$\mathbf{x} = \mathbf{Q}\mathbf{g} \tag{3.2-1}$$

where

$\mathbf{x}$ = state variables (n × 1)
$\mathbf{g}$ = control vector (m × 1)
$\mathbf{Q}$ = system matrix (n × m)
$\mathbf{d}$ = data vector (n × 1)

Following the idea of regularization, it is desired to find the **g** that minimizes the criteria

$$E = (\mathbf{d} - \mathbf{x}, \mathbf{d} - \mathbf{x}) + b(\mathbf{g}, \mathbf{g}) \tag{3.2-2}$$

where b is the regularization or smoothing parameter.

The solution is given by

$$\mathbf{g}^* = \left(\mathbf{Q}^T\mathbf{Q} + b\mathbf{I}\right)^{-1}\mathbf{Q}^T\mathbf{d} \tag{3.2-3}$$

and

$$\mathbf{x}^* = \mathbf{Q}\mathbf{g}^* \tag{3.2-4}$$

This can be written

$$\mathbf{x}^* = \mathbf{Q}\left(\mathbf{Q}^T\mathbf{Q} + b\mathbf{I}\right)^{-1}\mathbf{Q}^T\mathbf{d} \tag{3.2-5}$$

or

$$\mathbf{x}^* = \mathbf{A}\mathbf{d} \tag{3.2-6}$$

where

$$\mathbf{A} = \mathbf{Q}\left(\mathbf{Q}^T\mathbf{Q} + b\mathbf{I}\right)^{-1}\mathbf{Q}^T \tag{3.2-7}$$

It is important to note that **A** depends only on b and not directly on the data. **A** is sometimes referred to as the influence matrix.

Now consider a second problem that is the same as the original one except that the k*th* data point d_k is deleted. Also leave out the x_k variable but leave **g** the same size; that is, m × 1. This gives

$$\mathbf{z} - \mathbf{B}\mathbf{g} \tag{3.2-8}$$

where

$\mathbf{z}$ is $(n - 1 \times 1)$
$\mathbf{B}$ is $(n - 1 \times m)$ — the same as $\mathbf{Q}$ except that the k*th* row has been deleted
$\underline{\mathbf{d}}$ is $(n - 1 \times 1)$ — same as $\mathbf{d}$ except that d_k has been deleted

The new problem is to find the $\mathbf{g}$ that minimizes

$$E = (\underline{\mathbf{d}} - \mathbf{z}, \underline{\mathbf{d}} - \mathbf{z}) + b(\mathbf{g}, \mathbf{g}) \tag{3.2-9}$$

The solution is

$$\mathbf{g} = \left(\mathbf{B}^T\mathbf{B} + b\mathbf{I}\right)^{-1}\mathbf{B}^T\underline{\mathbf{d}} \tag{3.2-10}$$

$$\mathbf{z} = \mathbf{B}\mathbf{g} \tag{3.2-11}$$

Now let the deleted k*th* row of $\mathbf{Q}$ be denoted by the vector $\mathbf{q}$ $(m \times 1)$

$$\mathbf{q}^T = \begin{bmatrix} q_{k1} & q_{k2} & \cdots & q_{km} \end{bmatrix} \tag{3.2-12}$$

It is now possible to estimate the missing variable x_k using the above solution $\mathbf{g}$

$$x_{kp} = \mathbf{q}^T\mathbf{g} \tag{3.2-13}$$

Recall that this predicted value used all of the data except the deleted d_k*th* data point.

Now consider a third problem that will use the original model except that the predicted x_{kp} of the second problem will be used in place of d_k. We will now find the $\mathbf{g}$ that minimizes this problem ($\hat{\mathbf{g}}$). For this problem we will separate the variables so that the state variables are

$$\begin{bmatrix} z \\ x_k \end{bmatrix}$$

There is no loss of generality by placing x_k at the bottom. The data are

$$\begin{bmatrix} \underline{\mathbf{d}} \\ x_{kp} \end{bmatrix}$$

The system Equation 3.2-1 can be written in two parts

$$\mathbf{z} = \mathbf{B}\mathbf{g}$$

$$x_k = \mathbf{q}^T\mathbf{g}$$

The error criterion 3.2-2 becomes

$$E = (\underline{\mathbf{d}} - \mathbf{x}, \underline{\mathbf{d}} - \mathbf{z}) + \left(x_k - x_{kp}\right)^2 + b(\mathbf{g}, \mathbf{g}) \tag{3.2-14}$$

The minimum is given by

$$\left(\mathbf{B}^T\mathbf{B} + b\mathbf{I} + \mathbf{q}\mathbf{q}^T\right)\hat{\mathbf{g}} = \mathbf{B}\underline{\mathbf{d}} + \mathbf{q}x_{kp} \tag{3.2-15}$$

For ease of presentation, define the matrix **R** as

$$\mathbf{R} = \left(\mathbf{B}^T\mathbf{B} + b\mathbf{I}\right) \tag{3.2-16}$$

Using **R** and the Sherman Morrison formula, the inverse of the left-hand side of Equation 3.2-15 can be expressed as

$$\left(\mathbf{R} + \mathbf{q}\mathbf{q}^T\right)^{-1} = \mathbf{R}^{-1} - s\mathbf{R}^{-1}\mathbf{q}\mathbf{q}^T\mathbf{R}^{-1} \tag{3.2-17}$$

where the scalar s is given by

$$s = 1/\left(1 + \mathbf{q}^T\mathbf{R}^{-1}\mathbf{q}\right) \tag{3.2-18}$$

Substituting this into Equation 3.2-15 gives

$$\hat{\mathbf{g}} = \left(\mathbf{R}^{-1} - s\mathbf{R}^{-1}\mathbf{q}\mathbf{q}^T\mathbf{R}^{-1}\right)\left(\mathbf{B}^T\underline{\mathbf{d}} + \mathbf{q}x_{kp}\right) \tag{3.2-19}$$

Next, we are going to obtain a relationship between $\hat{\mathbf{g}}$ and $\mathbf{g}$ (the solution to the second problem). This is done by expanding Equation 3.2-19 and using Equation 3.2-10 wherever possible. This gives

$$\hat{\mathbf{g}} = \mathbf{g} - s\mathbf{R}^{-1}\mathbf{q}\mathbf{q}^T\mathbf{g} + \mathbf{R}^{-1}\mathbf{q}x_{kp} - s\mathbf{R}^{-1}\mathbf{q}\mathbf{q}^T\mathbf{R}^{-1}\mathbf{q}x_{kp} \qquad (3.2\text{-}20)$$

Recall that the predicted value x_{kp} is also related to $\mathbf{g}$ by Equation 3.2-13. When this is used in Equation 3.2-20 it can be shown that

$$\hat{\mathbf{g}} = \mathbf{g} \qquad (3.2\text{-}21)$$

This means that the solution to the third problem is the same as for the second problem.

Another way of expressing this result is that if the predicted value x_{kp} is used in place of d_k, the minimization returns x_{kp} as the solution. This remarkable fact will next be shown to lead to the cross-validation relationship.

Recall the influence matrix A of the original problem

$$\mathbf{x}^* = \mathbf{A}\mathbf{d} \qquad (3.2\text{-}6)$$

Now using the result that replacing the data point d_k with x_{kp} gives back x_{kp} as the solution

$$x_{kp} = \sum_{j \neq k} a_{kj} d_j + a_{kk} x_{kp} \qquad (3.2\text{-}22)$$

or

$$(1 - a_{kk}) x_{kp} = \sum_{j \neq k} a_{kj} d_j \qquad (3.2\text{-}23)$$

adding and subtracting $a_{kk}d_k$ from the right-hand side gives

$$(1 - a_{kk}) x_{kp} = \sum_{j=1} a_{kj} d_j - a_{kk} d_k \qquad (3.2\text{-}24)$$

where the summation now includes all of the terms. The final predicted value is

$$x_{kp} = \left(\sum_{j=1} a_{kj} d_j - a_{kk} d_k \right) \Big/ \left(1 - a_{kk}\right) \tag{3.2-25}$$

Since the original problem gives the solution $x_k^* = \sum_{j=1} a_{kj} d_j$ this can be written as

$$x_{kp} = \left(x_k^* - a_{kk} d_k\right) / \left(1 - a_{kk}\right) \tag{3.2-26}$$

Now recall our original goal of obtaining an expression for the difference between a predicted x_{kp} (without d_k) and the deleted point d_k. It can be shown that

$$x_{kp} - d_k = \left(x_k^* - d_k\right) / \left(1 - a_{kk}\right) \tag{3.2-27}$$

Finally repeating this for all of the points and summing the square of the terms gives the ordinary cross-validation formula $V_0(b)$ which is defined as

$$V_0(b) = \frac{1}{n} \sum_{k=1}^{n} \left(x_{kp} - d_k\right)^2 = \frac{1}{n} \sum_{k=1}^{n} \left(x_k^* - d_k\right)^2 / \left(1 - a_{kk}\right)^2 \tag{3.2-28}$$

Notice that the numerator is simply composed of the residuals of the original problem and that only the diagonal terms of **A** appear in the denominator.

In a paper by Golub, Heath, and Wahba (1979), the ordinary cross validation was extended to include the fact that a better estimation of b could be found if it minimized a general rotation of the transformation matrix **A**. This argument leads to the *generalized cross-validation* function

$$GVC(b) = \sum_{k=1}^{n} \left(x_k^* - d_k\right)^2 / \left(\mathrm{Tr}\left(\mathbf{I} - \mathbf{A}(b)\right)\right)^2 \tag{3.2-29}$$

where Tr denotes the trace of a matrix (the sum of the diagonals). It is important to note that only the diagonal terms of **A** are required to compute GCV.

The above static representation of the inverse problem is included only to explain the idea of cross validation. It is not a computationally feasible method for the following reason. Consider a fairly small dynamic inverse problem that has 10 measurements and 2 control variables per timestep and involves 200 timesteps. This amounts to a total of 2000 measurements and 400 unknown control variables. In the global static setting outlined above, the size of the matrix Q becomes (2000 × 400) and that of **A** (2000 × 2000), which are too large for any practical solution. In the next section we will show how to compute the diagonal terms of **A** without having to compute the rest of the matrix. This reduces the computational effort considerably. We will also show how the dynamic programming approach provides a natural setting for the application of generalized cross validation to the general inverse problem.

3.3 DYNAMIC PROGRAMMING AND GENERALIZED CROSS VALIDATION

For the general inverse problem discussed in Section 1.5, the least squares error was defined as

$$E(\mathbf{c},\mathbf{g}_j) = \sum_{j=1}^{N} \left(\mathbf{z}_j - \mathbf{d}_j, \mathbf{w}\left(\mathbf{z}_j - \mathbf{d}_j\right)\right) + \left(\mathbf{g}_j, \mathbf{B}\mathbf{g}_j\right) \qquad (3.3\text{-}1)$$

The weighting matrix **w** has been used in the first term to avoid confusion with the global matrix **A** of the preceding section. In this context, it is clear that the residual of the cross-validation problem of Section 3.2 is the first term of Equation 3.3-1 and that the second term is the smoothing part. This being the case, the global matrix **A** in Section 3.2 is the matrix that relates all of the measurements $\mathbf{d}_j$ to all of the variables $\mathbf{z}_j$. Recall that $\mathbf{z}_j$ is related to the state variables $\mathbf{x}_j$ by Equation 1.5-3 ($\mathbf{z}_j = \mathbf{Q}\mathbf{x}_j$). Considering that the $\mathbf{z}_j$'s and $\mathbf{d}_j$'s are vectors of length ($m_z \times 1$), then the global matrix **A** can be written in terms of submatrices $\mathbf{A}_{ik}$ ($m_z \times m_z$) as follows

$$\begin{aligned}
\mathbf{z}_1 &= \mathbf{A}_{11}\mathbf{d}_1 + \mathbf{A}_{12}\mathbf{d}_2 + \ldots + \mathbf{A}_{1N}\mathbf{d}_N \\
\mathbf{z}_2 &= \mathbf{A}_{21}\mathbf{d}_1 + \mathbf{A}_{22}\mathbf{d}_2 + \ldots + \mathbf{A}_{2N}\mathbf{d}_N \\
\ldots \quad & \ldots \quad \ldots \quad \ldots \quad \ldots \\
\mathbf{z}_N &= \mathbf{A}_{1N}\mathbf{d}_1 + \mathbf{A}_{2N}\mathbf{d}_2 + \ldots + \mathbf{A}_{NN}\mathbf{d}_N
\end{aligned} \tag{3.3-2}$$

We only need to find some way of computing the diagonal matrices $\mathbf{A}_{11}$,$\mathbf{A}_{22}$, etc. The trace of $\mathbf{A}$ is then easily computed as the sum of the diagonals of these submatrices and the generalized cross validation function (Equation 3.2-29) can then be evaluated.

Dohrmann, Trujillo, and Busby (1988) showed how the $\mathbf{A}_{ii}$'s could be computed as a natural extension of the dynamic programming formulas in Section 1.5. These relationships are considered important, and they will be derived in this section. First, define some new quantities as

$$\underline{\mathbf{M}}_{j+1} = \left(\mathbf{I} - \mathbf{P}\mathbf{D}_{j+1}\mathbf{H}_{j+1}\right)\mathbf{M} \tag{3.3-3}$$

$$\mathbf{E}_{j+1} = -\mathbf{P}\mathbf{D}_{j+1}\mathbf{P}^{T} \tag{3.3-4}$$

The equation for $\mathbf{s}_{n-1}$ (Equation 1.5-23, backward sweep) can be written

$$\mathbf{s}_n = -2\mathbf{Q}^{T}\mathbf{w}\mathbf{d}_n + \mathbf{M}_{n+1}^{T}\mathbf{s}_{n+1} \tag{3.3-5}$$

and the equation for the $\mathbf{x}_{j+1}$ (forward sweep) can be written as

$$\mathbf{x}_{j+1} = \underline{\mathbf{M}}_{j+1}\mathbf{x}_j + \mathbf{E}_{j+1}\mathbf{s}_{j+1} \tag{3.3-6}$$

In order to find a relationship between the vectors $\mathbf{s}_j$ and the data vectors $\mathbf{d}_j$ recall that the starting condition for $\mathbf{s}_N$ is given by

$$\mathbf{s}_N = -2\mathbf{Q}^{T}\mathbf{w}\mathbf{d}_N \tag{3.3-7}$$

Thus, obviously $\mathbf{s}_N$ depends only on $\mathbf{d}_N$. The next $\mathbf{s}_{N-1}$ can be written as

$$\mathbf{s}_{N-1} = -2\mathbf{Q}^T\mathbf{w}\mathbf{d}_{N-1} + \underline{\mathbf{M}}_N^T\mathbf{s}_N \qquad (3.3\text{-}8)$$

or

$$\mathbf{s}_{N-1} = -2\mathbf{Q}^T\mathbf{w}\mathbf{d}_{N-1} - 2\underline{\mathbf{M}}_N^T\mathbf{Q}^T\mathbf{w}\mathbf{d}_N \qquad (3.3\text{-}9)$$

so that $\mathbf{s}_{N-1}$ depends only on $\mathbf{d}_{N-1}$ and $\mathbf{d}_N$. In general we can write that

$$\mathbf{s}_n = \sum_{k=n} \mathbf{Y}_{nk}\mathbf{d}_k \qquad (3.3\text{-}10)$$

Notice that the index starts at k = n. Using this relationship in the recurrence formula for $\mathbf{s}_n$ (Equation 3.3-5) gives

$$\sum_{k=n}^{N} \mathbf{Y}_{nk}\mathbf{d}_k = -2\mathbf{Q}^T\mathbf{w}\mathbf{d}_n + \underline{\mathbf{M}}_{n+1}^T \sum_{k=n+1}^{N} \mathbf{Y}_{n+1,k}\mathbf{d}_k \qquad (3.3\text{-}11)$$

Since the data are arbitrary, we can equate coefficients to arrive at

$$\begin{aligned}
\mathbf{Y}_{nn} &= -2\mathbf{Q}^T\mathbf{w} \\
\mathbf{Y}_{n,n+1} &= \underline{\mathbf{M}}_{n+1}^T\mathbf{Y}_{n+1,n+1} \\
\mathbf{Y}_{n,n+2} &= \underline{\mathbf{M}}_{n+1}^T\mathbf{Y}_{n+1,n+2} \qquad (3.3\text{-}12) \\
\mathrm{K} & \qquad \mathrm{K} \\
\mathbf{Y}_{n.N} &= \underline{\mathbf{M}}_{n+1}^T\mathbf{Y}_{n+1,N}
\end{aligned}$$

This means that every term $\mathbf{Y}_{nn}$ will be equal to $-2\mathbf{Q}^T\mathbf{W}$ so that for the index n + 1

$$\mathbf{Y}_{n+1,n+1} = -2\mathbf{Q}^T\mathbf{w}$$

$$\mathbf{Y}_{n,n+1} = \underline{\mathbf{M}}_{n+1}^T\left(-2\mathbf{Q}^T\mathbf{w}\right)$$

and

$$\mathbf{Y}_{n+1,n+2} = \underline{\mathbf{M}}_{n+2}^T\left(-2\mathbf{Q}^T\mathbf{w}\right)$$

for the next index

$$\mathbf{Y}_{n,n+2} = \underline{\mathbf{M}}_{n+1}^T\underline{\mathbf{M}}_{n+1}^T\left(-2\mathbf{Q}^T\mathbf{w}\right)$$

In general the k_i*th* term is the product of the $\underline{\mathbf{M}}_k^T$ matrices starting with k + 1 and ending at i. This result is expressed as

$$\mathbf{Y}_{k,i} = \underline{\mathbf{M}}_{k+1}^T\underline{\mathbf{M}}_{k+2}^T \text{K } \underline{\mathbf{M}}_i^T\left(-2\mathbf{Q}^T\mathbf{w}\right) \qquad (3.3\text{-}13)$$

with

$$\mathbf{Y}_{k,k} = -2\mathbf{Q}^T\mathbf{w} \qquad (3.3\text{-}14)$$

The notation is somewhat cumbersome but will prove to be extremely useful in the next derivation. As an example, consider the computation of $\mathbf{Y}_{2,6}$. It is given as

$$\mathbf{Y}_{2,6} = \underline{\mathbf{M}}_3^T\underline{\mathbf{M}}_4^T\underline{\mathbf{M}}_5^T\underline{\mathbf{M}}_6^T\left(-2\mathbf{Q}^T\mathbf{w}\right)$$

The next step is to derive a relationship between $\mathbf{x}_i$'s and the $\mathbf{d}_i$'s. To do this we will start at the beginning with $\mathbf{x}_1$. Recall that the optimal initial condition for the first state $\mathbf{x}_1$ was given by Equation 1.5.1-2 as

$$\mathbf{x}_1 = -\mathbf{R}_1^{-1}\,\mathbf{s}_1/2$$

Using Equation 3.3-10 for $\mathbf{s}_1$ gives

$$\mathbf{x}_1 = -\mathbf{R}_1^{-1}\left(\mathbf{Y}_{11}\mathbf{d}_1 + \mathbf{Y}_{12}\mathbf{d}_2 + \ldots + \mathbf{Y}_{1N}\mathbf{d}_N\right)/2 \tag{3.3-15}$$

which is a linear relationship between $\mathbf{x}_1$ and all of the data. Now to obtain a relationship between the $\mathbf{z}_1$ and the data, use $\mathbf{z}_1 = \mathbf{Q}\mathbf{x}_1$ to give

$$\mathbf{z}_1 = -\mathbf{Q}\mathbf{R}_1^{-1}\left(\mathbf{Y}_{11}\mathbf{d}_1 + \mathbf{Y}_{12}\mathbf{d}_2 + \ldots + \mathbf{Y}_{1N}\mathbf{d}_N\right)/2 \tag{3.3-16}$$

It is clear that the submatrix $\mathbf{A}_{11}$ from Equation 3.3-2 can be identified as the first term of the above equation $-\mathbf{Q}\mathbf{R}_1^{-1}\mathbf{Y}_{11}/2$ and since $\mathbf{Y}_{11} = -2\mathbf{Q}^T\mathbf{w}$ this gives

$$\mathbf{A}_{11} = \mathbf{Q}\mathbf{R}_1^{-1}\mathbf{Q}^T\mathbf{w} \tag{3.3-17}$$

This is the first submatrix of the global matrix **A** (Equation 3.3-2).

Let us compute the second state using Equation 3.3-6

$$\mathbf{x}_2 = \underline{\mathbf{M}}_2\mathbf{x}_1 + \mathbf{E}_2\mathbf{s}_2 \tag{3.3-18}$$

or

$$\begin{aligned}\mathbf{x}_2 = &-\underline{\mathbf{M}}_2\mathbf{R}_1^{-1}\left(\mathbf{Y}_{11}\mathbf{d}_1 + \mathbf{Y}_{12}\mathbf{d}_2 + \ldots + \mathbf{Y}_{1N}\mathbf{d}_N\right)/2 \\ &+ \mathbf{E}_2\left(\mathbf{Y}_{22}\mathbf{d}_2 + \mathbf{Y}_{23}\mathbf{d}_3 + \ldots + \mathbf{Y}_{2N}\mathbf{d}_N\right)\end{aligned} \tag{3.3-19}$$

We are only interested in the submatrix ($\mathbf{S}_L$) that multiples $\mathbf{d}_2$ it is given by

$$\mathbf{S}_L = -\underline{\mathbf{M}}_2\mathbf{R}_1^{-1}\mathbf{Y}_{12}/2 + \mathbf{E}_2\mathbf{Y}_{22}$$

or by eliminating the $\mathbf{Y}_{12}$ and $\mathbf{Y}_{22}$ we get

$$\mathbf{S}_L = \left(\underline{\mathbf{M}}_2\mathbf{R}_1^{-1}\underline{\mathbf{M}}_2^T - 2\mathbf{E}_2\right)\left(\mathbf{Q}^T\mathbf{w}\right) \tag{3.3-20}$$

and the second submatrix is now computed as

$$\mathbf{A}_{22} = \mathbf{Q}\left(\underline{\mathbf{M}}_2\mathbf{R}_1^{-1}\underline{\mathbf{M}}_2^T - 2\mathbf{E}_2\right)\left(\mathbf{Q}^T\mathbf{w}\right) \tag{3.3-21}$$

It is now possible to obtain a recursive formula for some submatrices $\mathbf{L}_{ii}$ by first defining

$$\mathbf{L}_{11} = \mathbf{R}_1^{-1} \tag{3.3-22}$$

The second matrix is

$$\mathbf{L}_{22} = \underline{\mathbf{M}}_2 \mathbf{R}_1^{-1} \underline{\mathbf{M}}_2^T - 2\mathbf{E}_2$$

or

$$\mathbf{L}_{22} = \underline{\mathbf{M}}_2 \mathbf{L}_{11} \underline{\mathbf{M}}_2^T - 2\mathbf{E}_2 \tag{3.3-23}$$

By assuming a form

$$\mathbf{x}_j = \mathbf{L}_{ji} \mathbf{Q}^T \mathbf{w} \mathbf{d}_i$$

it can easily be shown that the recursive formula for $\mathbf{L}_{k,k}$ is

$$\mathbf{L}_{k+1,k+1} = \underline{\mathbf{M}}_{k+1} \mathbf{L}_{kk} \underline{\mathbf{M}}_{k+1}^T - 2\mathbf{E}_{k+1} \tag{3.3-24}$$

Each submatrix $\mathbf{A}_{kk}$ is then calculated using

$$\mathbf{A}_{kk} = \mathbf{Q} \mathbf{L}_{kk} \mathbf{Q}^T \mathbf{w} \tag{3.3-25}$$

The trace is then the sum of the diagonals of $\mathbf{A}_{kk}$.

It should be emphasized that the above equations represent an enormous computational savings. Instead of having to calculate the entire matrix **A**, we have reduced the problem to dealing with the submatrices $\mathbf{L}_{kk}$ which are of size m × m. Consider a practical thermal problem where m = 400, $m_z = 10$, and there are N = 500 steps. The size of the global matrix **A** is a function of the measurements m_z and the total number of steps, in this case (5000 × 5000). In contrast, the submatrices $\mathbf{L}_{kk}$ are 400 × 400, which is still fairly large yet manageable. Thus, generalized cross validation is a practical partner to dynamic programming in solving the general inverse problem.

There are some practical computational notes that should be mentioned. In general, one should always be aware of having to compute or store full matrices and avoid unnecessary calculations. For example, the matrix $\mathbf{E}_k$ is never formed but is replaced in the computations by $-\mathbf{P}\mathbf{D}_k\mathbf{P}^T$. Thus only the smaller matrices $\mathbf{D}_k$ ($m_g \times m_g$) need to be stored during the backward sweep. The same idea applies to the matrices $\mathbf{M}_k$. It appears that the full matrices $\mathbf{L}_{kk}$ cannot be avoided.

Exercise

The above derivation used the optimal initial conditions given by Equation 1.5.1-2. A more common case is where the initial conditions are assumed to be known. Rederive the above equations for the trace of **A** for this case.

3.4 CHANDRASEKHAR EQUATIONS

The dynamic programming recursive formulas in Section 1.5 were developed in terms of full symmetric matrices $\mathbf{R}_n$ ($m \times m$). In this section we will present some alternate formulas that will use matrices only of order ($m \times m_z$) and ($m \times m_g$). These formulas are essential if we are to analyze practical systems of m equal to or greater than 1000. It is frequently the case that m_g and m_z are much less than m. For example, a finite element heat conduction model can easily have 500 nodes and yet have only 10 thermocouple measurements and 7 unknown heat fluxes distributed around the model. Thus, instead of having to deal with matrices of order (500×500), we will only need to concern ourselves with matrices of order (500×10) and (500×7). Besides the reduction in storage, the reduction in computational effort is at least two orders of magnitude. The main restriction is that the system matrices **B**, **A**, **P**, and **M** must be constant. This is not a serious restriction, and in Chapter 7 we will show how to deal with nonlinear systems using quasilinear techniques. Chapter 2 showed how to represent the system matrix **M** with several approximate operators that retain the sparseness of the original matrix. In the following derivation, **M** and **P** should be thought of as operators.

The main idea in this section is to derive recurrence relationships for incremental values of $\mathbf{R}_n$. That is, we are interested in quantities such as $\delta\mathbf{R}_n = \mathbf{R}_n - \mathbf{R}_{n-1}$ (note: δ represents the incremental quantity) and how they relate to $\delta\mathbf{R}_{n-1}$. These formulas are also known as the Chandrasekhar equations (see Morf, 1974; Budgell, 1987).

Let us first develop the incremental formulas for $\delta\mathbf{R}_n$. In a later section we will complete the formulas for the $\mathbf{s}_n$'s. Recall the formulas for $\mathbf{R}_n$

$$\mathbf{D}_n^{-1} = 2\mathbf{B} + 2\mathbf{P}^T\mathbf{R}_n\mathbf{P} \tag{3.4-1}$$

$$\mathbf{R}_{n-1} = \mathbf{Q}^T\mathbf{A}\mathbf{Q} + \mathbf{M}^T\left(\mathbf{R}_n - \mathbf{R}_n\mathbf{P}\mathbf{D}_n 2\mathbf{P}^T\mathbf{R}_n\right)\mathbf{M} \tag{3.4-2}$$

The intermediate variable $\mathbf{H}_n$ has been replaced with Equation 1.5-16. The matrix $\mathbf{K}_n$ is now defined as

$$\mathbf{K}_n = \mathbf{R}_n\mathbf{P}\mathbf{D}_n \tag{3.4-3}$$

The formulas will now be expressed in terms of $\mathbf{K}_n$, which is of order $(m \times m_g)$. In addition, define the increments

$$\delta\mathbf{R}_n = \mathbf{R}_n - \mathbf{R}_{n-1} \tag{3.4-4}$$

$$\delta\mathbf{D}_n^{-1} = \mathbf{D}_n^{-1} - \mathbf{D}_{n-1}^{-1} \tag{3.4-5}$$

Since $\mathbf{B}$ and $\mathbf{P}$ are constant, these equations lead to the incremental relationship

$$\delta\mathbf{D}_n^{-1} = 2\mathbf{P}^T\delta\mathbf{R}_n\mathbf{P} \tag{3.4-6}$$

Using $\mathbf{K}_n$ in Equation 3.4-2 gives

$$\mathbf{R}_{n-1} = \mathbf{Q}^T\mathbf{A}\mathbf{Q} + \mathbf{M}^T\left(\mathbf{R}_n - 2\mathbf{K}_n\mathbf{D}_n^{-1}\mathbf{K}_n^T\right)\mathbf{M} \tag{3.4-7}$$

In incremental form, this can be written as

$$\delta\mathbf{R}_{n-1} = \mathbf{M}^{T}\left(\delta\mathbf{R}_{n} - 2\mathbf{K}_{n}\mathbf{D}_{n}^{-1}\mathbf{K}_{n}^{T} + 2\mathbf{K}_{n-1}\mathbf{D}_{n-1}^{-1}\mathbf{K}_{n-1}^{T}\right)\mathbf{M} \qquad (3.4\text{-}8)$$

We will also require an incremental formula for $\mathbf{K}_n$. It can easily be shown that

$$\mathbf{K}_{n-1} = \mathbf{K}_{n} + \left(2\mathbf{K}_{n}\mathbf{P}^{T}\delta\mathbf{R}_{n}\mathbf{P} - \delta\mathbf{R}_{n}\mathbf{P}\right)\mathbf{D}_{n-1} \qquad (3.4\text{-}9)$$

Using this equation in Equation 3.4-8 for $\mathbf{K}_{n-1}$ and after some rearranging, it can be shown that

$$\begin{aligned}\delta\mathbf{R}_{n-1} &= \mathbf{M}^{T}\left[\mathbf{I} - 2\mathbf{K}_{n}\mathbf{P}^{T}\right]\left(\delta\mathbf{R}_{n} + 2\delta\mathbf{R}_{n}\mathbf{P}\mathbf{D}_{n-1}\mathbf{P}^{T}\delta\mathbf{R}_{n}\right) \\ &\times\left[\mathbf{I} - 2\mathbf{K}_{n}\mathbf{P}^{T}\right]^{T}\mathbf{M}\end{aligned} \qquad (3.4\text{-}10)$$

This incremental formula is now in a form that lends itself to recurrence formulas in terms of $\mathbf{Y}_k$ and $\mathbf{L}_k$, which are defined as

$$\delta\mathbf{R}_{n} = \mathbf{Y}_{n}\mathbf{L}_{n}\mathbf{Y}_{n}^{T} \qquad (3.4\text{-}11)$$

Using this definition in Equation 3.4-10 and equating like terms gives

$$\mathbf{Y}_{n-1} = \mathbf{M}^{T}\left[\mathbf{I} - 2\mathbf{K}_{n}\mathbf{P}^{T}\right]\mathbf{Y}_{n} \qquad (3.4\text{-}12)$$

$$\mathbf{L}_{n-1} = \mathbf{L}_{n} + \mathbf{L}_{n}\mathbf{Y}_{n}^{T}2\mathbf{P}\mathbf{D}_{n-1}\mathbf{P}^{T}\mathbf{Y}_{n}\mathbf{L}_{n} \qquad (3.4\text{-}13)$$

The initial conditions for the above matrices are obtained by expanding the first two terms of the backward sweep, which are given by

$$\mathbf{R}_{N} = \mathbf{Q}^{T}\mathbf{A}\mathbf{Q}$$

$$\mathbf{R}_{N-1} = \mathbf{Q}^{T}\mathbf{A}\mathbf{Q} + \mathbf{M}^{T}\left(\mathbf{Q}^{T}\mathbf{A}\mathbf{Q} - \mathbf{Q}^{T}\mathbf{A}\mathbf{Q}2\mathbf{P}\mathbf{D}_{N}\mathbf{P}^{T}\mathbf{Q}^{T}\mathbf{A}\mathbf{Q}\right)\mathbf{M}$$

The first increment is then given as

$$\delta \mathbf{R}_N = \mathbf{R}_N - \mathbf{R}_{N-1} = \mathbf{M}^T\mathbf{Q}^T\left(-\mathbf{A} + \mathbf{A}\mathbf{Q}2\mathbf{P}\mathbf{D}_N\mathbf{P}^T\mathbf{Q}^T\mathbf{A}\right)\mathbf{Q}\mathbf{M} \quad (3.4\text{-}14)$$

Since

$$\delta \mathbf{R}_N = \mathbf{Y}_N\mathbf{L}_N\mathbf{Y}_N^T \quad (3.4\text{-}15)$$

this gives

$$\mathbf{Y}_N = \mathbf{M}^T\mathbf{Q}^T \quad (3.4\text{-}16)$$

$$\mathbf{L}_N = -\mathbf{A} + \mathbf{A}\mathbf{Q}2\mathbf{P}\mathbf{D}_N\mathbf{P}^T\mathbf{Q}^T\mathbf{A} \quad (3.4\text{-}17)$$

where $\mathbf{Y}_N$ is of order $(m \times m_z)$ and $\mathbf{L}_N$ is of order $(m_z \times m_z)$. The initial condition for $\mathbf{D}_N$ is

$$\mathbf{D}_N^{-1} = 2\mathbf{B} + 2\mathbf{P}^T\mathbf{Q}^T\mathbf{A}\mathbf{Q}\mathbf{P} \quad (3.4\text{-}18)$$

These equations can be used in the recurrence equation for $\mathbf{s}_n$ which is rewritten in terms of $\mathbf{K}_n$

$$\mathbf{s}_{n-1} = -2\mathbf{Q}^T\mathbf{A}\mathbf{d}_{n-1} + \mathbf{M}^T\left(\mathbf{I} - 2\mathbf{K}_n\mathbf{P}^T\right)\mathbf{s}_n \quad (3.4\text{-}19)$$

The forward sweep is also expressed in terms of $\mathbf{K}_n$

$$\mathbf{g}_{n-1} = -\mathbf{D}_n\mathbf{P}^T\mathbf{s}_n - 2\mathbf{K}_n^T\mathbf{M}\mathbf{c} \quad (3.4\text{-}20)$$

The complete sequence of operations is organized as follows:

a. The initial conditions

$$\mathbf{D}_N^{-1} = 2\mathbf{B} + 2\mathbf{P}^T\mathbf{Q}^T\mathbf{A}\mathbf{Q}\mathbf{P}$$

$$\mathbf{K}_N = \mathbf{Q}^T\mathbf{A}\mathbf{Q}\mathbf{P}\mathbf{D}_N$$

$$\mathbf{Y}_N = \mathbf{M}^T\mathbf{Q}^T$$

$$\mathbf{L}_N = -\mathbf{A} + \mathbf{A}\mathbf{Q}2\mathbf{P}\mathbf{D}_N\mathbf{P}^T\mathbf{Q}^T\mathbf{A}$$

$$\mathbf{s}_N = -2\mathbf{Q}^T\mathbf{A}\mathbf{d}_N$$

b. The backward sweep

$$\mathbf{D}_{n-1}^{-1} = \mathbf{D}_{n}^{-1} - 2\mathbf{P}^{T}\mathbf{Y}_{n}\mathbf{L}_{n}\mathbf{Y}_{n}^{T}\mathbf{P}$$

$$\mathbf{K}_{n-1} = \mathbf{K}_{n} + \left(2\mathbf{K}_{n}\mathbf{P}^{T}\mathbf{Y}\mathbf{L}_{n}\mathbf{Y}_{n}^{T}\mathbf{P} - \mathbf{Y}_{n}\mathbf{L}_{n}\mathbf{Y}_{n}^{T}\mathbf{P}\right)\mathbf{D}_{n-1}$$

$$\mathbf{Y}_{n-1} = \mathbf{M}^{T}\left[\mathbf{I} - 2\mathbf{K}_{n}\mathbf{P}^{T}\right]\mathbf{Y}_{n}$$

$$\mathbf{L}_{n-1} = \mathbf{L}_{n} + \mathbf{L}_{n}\mathbf{Y}_{n}^{T}2\mathbf{P}\mathbf{D}_{n-1}\mathbf{P}^{T}\mathbf{Y}_{n}\mathbf{L}_{n}$$

$$\mathbf{s}_{n-1} = -2\mathbf{Q}^{T}\mathbf{A}\mathbf{d}_{n-1} + \mathbf{M}^{T}\left[\mathbf{I} - 2\mathbf{K}_{n}\mathbf{P}^{T}\right]\mathbf{s}_{n}$$

c. The forward sweep

$$\mathbf{g}_{j} = -\mathbf{D}_{j+1}\mathbf{P}^{T}\mathbf{s}_{j+1} - 2\mathbf{K}_{j+1}^{T}\mathbf{M}\mathbf{x}_{j}$$

$$\mathbf{x}_{j+1} = \mathbf{M}\mathbf{x}_{j} + \mathbf{P}\mathbf{g}_{j}$$

These formulas have succeeded in computing the backward and forward sweeps with smaller matrices. It is clear that much care must be taken in performing the above calculations to avoid repeating similar calculations and to minimize storage requirements. The sizes of the above matrices are summarized below:

Matrix	**Size**
$\mathbf{D}_n$	$(m_g \times m_g)$
$\mathbf{K}_n$	$(m \times m_g)$
$\mathbf{Y}_n$	$(m \times m_z)$
$\mathbf{L}_n$	$(m_z \times m_z)$
$\mathbf{P}$	$(m \times m_g)$
$\mathbf{M}$	$(m \times m)$

Since this formulation does not explicitly compute the matrix $\mathbf{R}_n$, it cannot be used to determine the optimal initial conditions given by Equation 1.5.1-2. This is not very important because this formulation is primarily intended for very large systems where it will be assumed that the initial conditions are known to a greater degree of certainty than the other unknowns.

Exercise

1. Verify Equation 3.4-10.
2. It is interesting to note that determining the optimal order in which a group of matrices is multiplied can be easily solved using dynamic programming. Consider a group of matrices K_i that are to be multiplied

$$(K_1)\times(K_2)\times\ldots\times(K_n)$$

Each matrix has r_i rows and c_i columns, and the multiplication of matrices i and j requires $r_i c_i r_i$ multiplications. Determine the order of matrix multiplication to minimize the total number of multiplications (see Aho, 1974).

Chapter 4

The Inverse Heat Conduction Problem

4.1 INTRODUCTION

Most heat conduction problems are concerned with the determination of temperatures at interior points when certain initial and boundary conditions are given, such as temperature or heat flux. These problems are termed "direct" problems because the solution involves a direct integration of differential equations with known initial conditions and known forcing terms. On the other hand, the inverse problem is concerned with the estimation of the applied heat fluxes based on measured temperature data. Typically, these problems arise because measurements can only be made in easily accessible locations, or perhaps a desired variable can only be measured indirectly. For example, a temperature history measurement is made on the outside of a pressure vessel, and we want to estimate the heat flux history on the inside surface.

The inverse heat conduction problem is one of the more difficult problems to solve for two main reasons. First, in the direct problem, the high-frequency components of the applied heat flux are damped as the heat flow diffuses through the solid medium. In the inverse problem, the opposite occurs. The high-frequency components or noise in the measurements will be amplified in the projection to the surface, and the resulting surface condition estimations can be easily overwhelmed by the noise in the interior measurement. Second, the physics of heat conduction introduces a natural lag between the applied heat flux and the temperature response away from the flux. Thus, a step change in the surface heat flux will not be fully felt in the interior until a finite amount of time has passed.

The inverse heat conduction problem has been the subject of considerable research. Several methods have been used to solve the inverse problem. Some of these include graphical (Stolz, 1960), polynomial (Frank, 1963; Mulholland et al., 1975; Arledge and Haji-Sheikh, 1977), Laplace transform methods (Krzysztof et al., 1981), and exact methods (Deverall and Channapragada, 1966; Burggraf, 1964; Sparrow et al., 1975). Most of these methods can only be used with precise data that are continuously differentiable. Meric (1979) used a combination of finite elements and conjugate gradient methods to solve a nonlinear control problem. Beck and others (1982, 1985) have successfully used methods based on sensitivity coefficients to solve problems in one and two dimensions.

In this chapter we will apply dynamic programming and generalized cross validation to the inverse heat conduction problem. These methods offer a great deal of flexibility in the type of model, the number and location of the measurements, and the number and location of the unknown heat fluxes. In addition, with the application of the Chandrasekhar equations, very large models can be easily accommodated, and in a later chapter we will show how to extend these concepts to nonlinear problems. Much of this material is the result of research performed by the authors (Busby and Trujillo, 1985b; Trujillo, 1978; Trujillo and Wallis, 1989).

4.2 ONE-DIMENSIONAL EXAMPLE

The simplest example of an inverse heat conduction problem is a one-dimensional rod where the spatial derivatives are represented with a finite difference approximation. The unknown heat flux Q(t) will be applied at one end, and the measurement will be taken at the other end, which will be assumed to be insulated. The model represents a solid rod of length 1.0 divided into 10 equal segments of length 0.10. There are a total of 11 nodes, and unit thermal properties are used. The semi-discrete differential equations for the temperatures of the nodes are given by:

$$0.05\dot{T}_1 = 10(T_2 - T_1) + Q(t)$$

$$0.10\dot{T}_2 = 10(T_1 - T_2) + 10(T_3 - T_2)$$

$$\ldots \quad \ldots \quad \ldots \quad \ldots$$

$$0.05\dot{T}_{11} = 10(T_{10} - T_{11})$$

In matrix-vector form these equations become

$$\mathbf{C}\dot{\mathbf{T}} = \mathbf{KT} + \mathbf{Pq} \tag{4.2-1}$$

These differential equations can be replaced with any one of several approximations (see Section 2.4) to give

$$\mathbf{T}_{j+1} = \mathbf{MT}_j + \mathbf{Pq}_j \tag{4.2-2}$$

In this section the exponential matrix will be used, although it is more common to use the Crank–Nicolson approximation for the heat conduction equation, especially for large systems with sparse matrices.

In lieu of real experimental data it is possible to apply known heat fluxes to a model and generate simulated measurements by directly integrating the equations. Since the true heat fluxes are known, the performance of the inverse methods can be measured directly. It is also possible to simulate noise in a measurement and ascertain its effect on the inverse estimations. Figure 4.1 shows such an applied triangular heat flux and the subsequent temperature response of node 11, which represents a thermocouple at the insulated surface. It can be seen that the thermocouple response lags behind the applied heat. The inverse problem is to estimate the applied heat flux history given only the thermocouple measurement and the thermal model. This will be done using the dynamic programming methods outlined in the previous chapters.

If the simulated measurement is used without adding noise, the inverse estimation of the heat flux is *indistinguishable* from

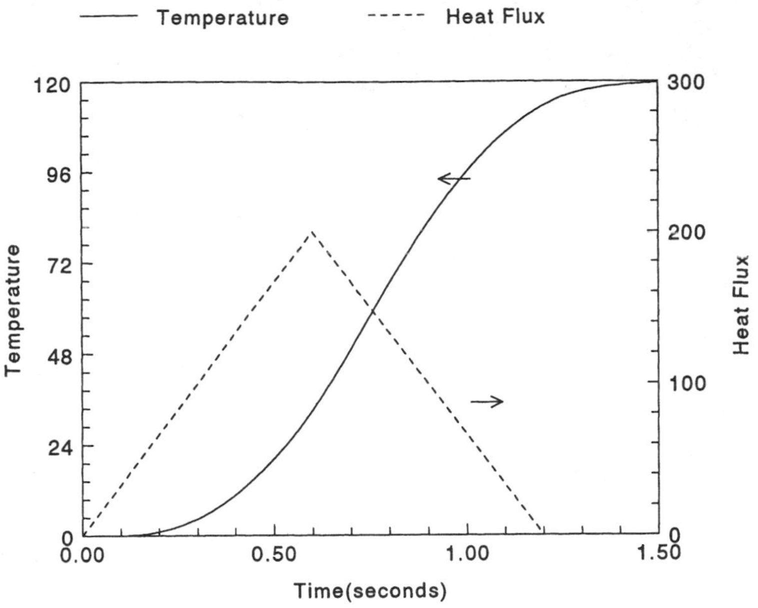

FIGURE 4.1
Heat flux and temperature response.

the actual one. The more realistic case is when noise is added to the thermocouple measurement. The noise was generated using a computer routine that generates a normally distributed variable with a given variance (sigma) (Press et al., 1990). In the first case, a sigma of 0.10 was used. The inverse problem was then solved with a smoothing value of b = 3.4E–05. The estimation of the heat flux is shown in Figure 4.2. It can be seen that the noise has induced slight oscillations in the estimated heat flux; however, the overall agreement is excellent. In a strict sense, the heat flux obtained from the inverse solution is constant over each time interval. However, for plotting purposes, it is represented with straight lines. In order to further ascertain the effects of noise, a second case was investigated using a sigma of 0.25. The results are shown in Figure 4.3, which shows that the larger noise level has started to degrade the estimation of the heat flux. A smoothing value of b = 8.8E–05 was used in the inverse solution. In a later section, we will show how to obtain smoother estimates of the heat flux by using a first-order regularization technique. This technique simply regulates the first derivative of the heat flux instead of the heat flux itself.

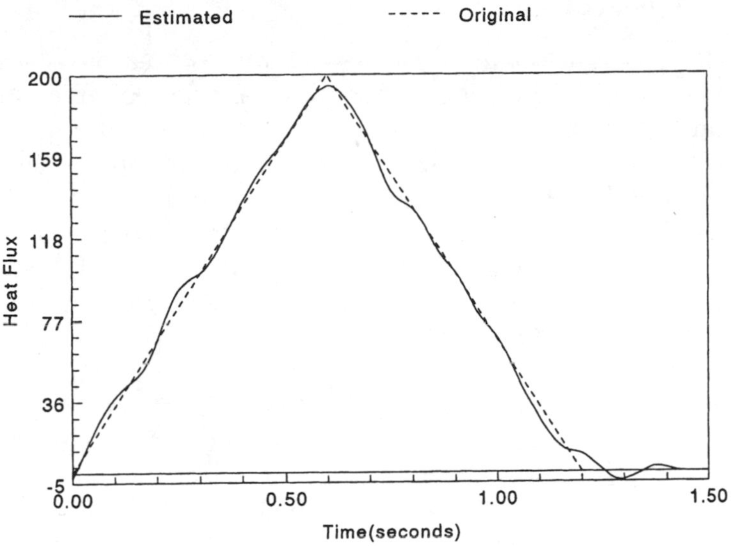

FIGURE 4.2
Comparison of estimated and original heat flux (noise = 0.10).

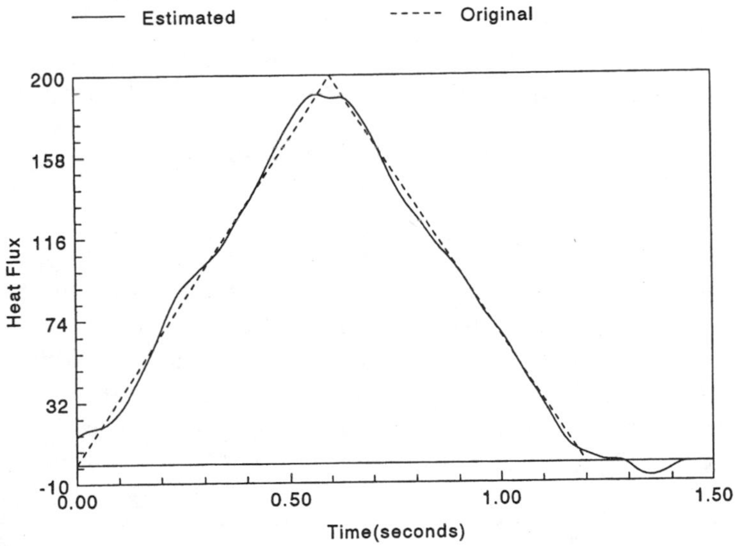

FIGURE 4.3
Comparison of estimated and original heat flux (noise = 0.25).

4.2.1 Quenching Curves

The previous example dealt with the estimation of a heat flux versus time. There is another important class of inverse heat conduction problems that involve the estimation of a heat flux versus a surface temperature. This cross correlation is especially useful in boiling heat transfer where the heat flux can undergo a large change while the surface temperature undergoes only a relatively small change. The difficulty with such problems is that an estimate (heat flux) is to be plotted against another estimate (surface temperature). A quenching operation is one such process.

In order to demonstrate the ability of the inverse method to estimate these quenching curves, a one-dimensional quench was numerically simulated using a direct heat transfer process. The slab was 2.0 inches thick and was quenched on both faces so that only 1.0 inch needed to be modeled. The thermocouple was placed in the middle of the slab. In a real problem, the thermocouple is usually placed closer to the surface. The center location was chosen to provide a more severe test of the inverse method. The thermal properties approximate those of steel and are:

$$\text{thermal conductivity} = 2.59\text{E} - 04 \text{ BTU/(inch-second-F)}$$

$$\text{specific heat} = 0.12 \text{ BTU/(lbm-F)}$$

$$\text{density} = 0.289 \text{ lbm/in}^3$$

The heat flux versus temperature curve is shown in Figure 4.4. This simplified curve was used as input to a direct heat transfer program to generate the temperature history at the center of the slab. The slab was at an initial uniform temperature of 1900°F. An integration timestep of 0.25 seconds was used in the simulation, but the results were only sampled at 1.0-second intervals. Also, the temperatures were only recorded to one digit after the decimal point, i.e. 1832.3.

The results of the inverse problem are shown in Figure 4.4 for a smoothing parameter of b = 10. As expected, the cross-correlation plots accentuate any oscillations in the estimations. Overall, the estimate gives excellent agreement except at the lower temperatures. This is because the data were truncated after 150 seconds. The main point of interest is to ascertain whether the inverse method could estimate the very sharp drop in the

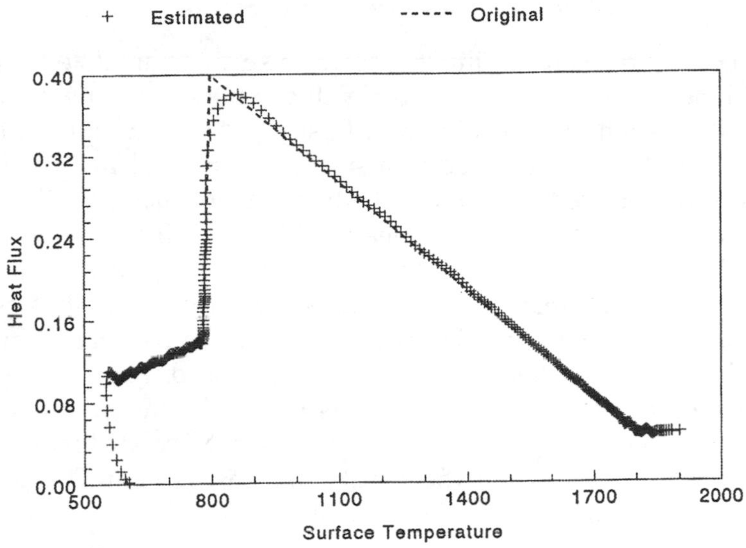

FIGURE 4.4
Comparison of estimated and original heat flux correlation.

heat flux that occurs between 780°F and 800°F. This type of behavior is common in boiling heat transfer where the boiling is changing character from film boiling to nucleate boiling and finally to natural convection. A more realistic model will have to include the temperature variation in the thermal properties since the metal undergoes a very large change in temperature. These nonlinear inverse problems will be discussed in a later chapter.

These numerical simulations are useful in validating the ability of inverse methods to generate accurate estimates of the unknown heat fluxes. They can also prove useful in practical problems, especially those involving complex geometries, nonlinear properties, and several unknown heat fluxes. Although the *actual* heat fluxes are unknown, something is usually known about them, such as their order of magnitude or perhaps a rough estimate of a film coefficient. Using this knowledge, one can *a priori* simulate a process using the direct solution and experiment with such items as the number and locations of the thermocouples and the sampling interval. Also, the numerical data can be contaminated with various amounts of noise to better simulate real data. By observing the effects of these variables on the inverse estimations, one can better design the actual experiments.

4.2.2 Generalized Cross Validation Example

This next example illustrates the use of generalized cross validation to estimate the optimal value of the smoothing parameter. It is taken from Trujillo and Busby (1989) and represents a semi-infinite slab subjected to a surface heat flux. This problem was also analyzed by Beck and Murio (1986) using a different method of solution. The true heat flux is a unit flux between 0.2 and 0.6 seconds. The rest of the time it is zero. The semi-infinite slab was approximated with a finite-difference model that used a total of 41 nodes spaced 0.10 units apart. The model is dimensionless and unit thermal properties were used. An integration time increment of 0.01 seconds was used. The measurement is located at a distance of 1.0 units from the surface. A direct solution was used to generate the simulated measurement. This problem is particularly difficult because the heat flux is a step function which contains a great deal of high frequency information that will be damped by the time it reaches the measurement location. For this reason it provides an excellent example with which to test any inverse method.

In order to simulate real measurements, a random noise level of $\sigma = 0.002$ was also added to the temperatures. The resulting noisy data are shown in Figure 4.5. Since the true heat flux is known for this example, it is possible to compute an error between the estimated heat flux q_i and the true heat flux $\hat{q}_i$. This error is S and it depends on the smoothing parameter b:

$$S^2(b) = \frac{1}{N}\sum_{i=1}^{N}\left(q_i - \hat{q}_i\right)^2 \tag{4.2.2-1}$$

With this function we can evaluate the performance of the generalized cross validation function V(b). This was done by solving the inverse problem for a series of smoothing parameters. The results are shown in Figure 4.6, which illustrates the usefulness of the GCV method. Since the true heat flux is known, one would select b to minimize S(b). However, in a real situation, one can only minimize V(b). Figure 4.6 illustrates the fact that the minimum of V(b) gives an excellent estimate to the optimal regularization parameter given by the minimum of S(b). The estimated heat flux obtained by using the optimal b is shown in Figure 4.7 together with the true heat flux. Although the estimated heat

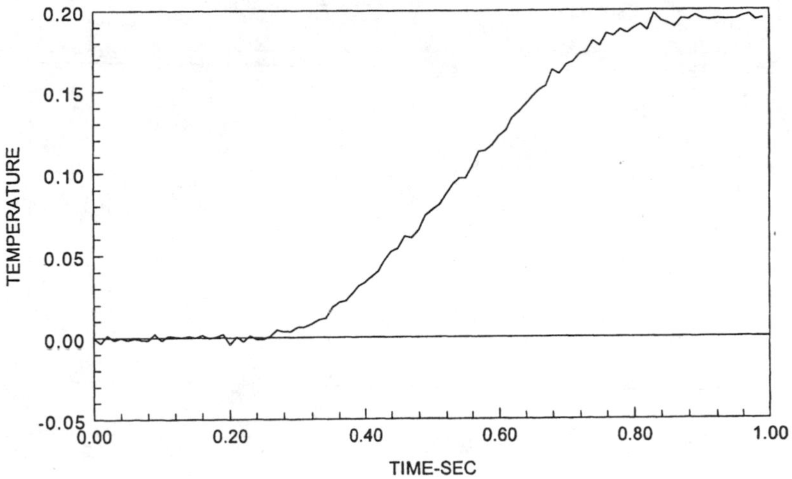

FIGURE 4.5
Temperature data with added noise.

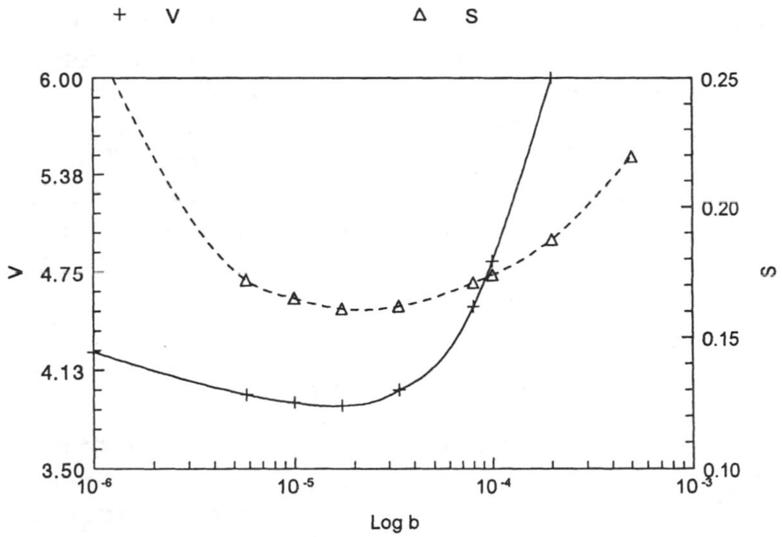

FIGURE 4.6
Variation of S and V with the smoothing parameter b.

flux is smooth, it does not match the true heat flux very well and appears to represent a typical least squares solution. This is undoubtedly due to the absence of the high frequency content in the measurement and to the least squares formulation of the inverse problem. However, the results of Figure 4.7 lead to the

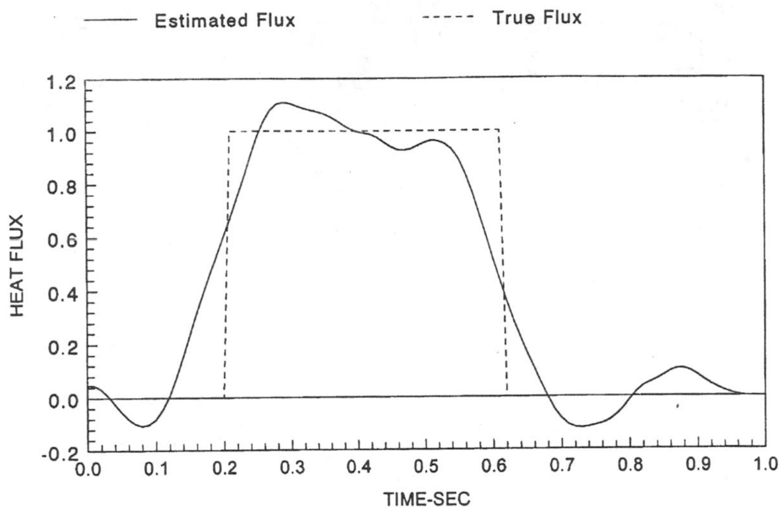

FIGURE 4.7
Comparison of estimated and original heat flux.

conclusion that this is the closest we can get to the true heat flux with the current model. At this point, if one desired more resolution, one could relocate the thermocouple closer to the surface and repeat the experiment. The advantage of using GCV is that it will eliminate one parameter from the problem. That is, one can be reasonably sure that the optimal smoothing parameter is being used. As the noise level increases, the GCV estimate of b is not as good. See Trujillo and Busby (1989) for more details.

4.3 TWO-DIMENSIONAL EXAMPLE

This two-dimensional example will illustrate the application of the inverse methods and generalized cross validation (GCV) to a model involving two measurements and two unknown heat fluxes (Trujillo and Busby, 1989). The model is based on a finite element approximation to a slab shown in Figure 4.8. The model consists of 25 nodes and 16 quadrilateral elements. Unit thermal properties were used. The heat fluxes were applied to the top and left side, with the other two sides insulated. The measurements were located at the midpoints of the insulated sides, nodes 11 and 23. A time increment of 0.01 units was used. The measurements were simulated by solving a direct problem with the

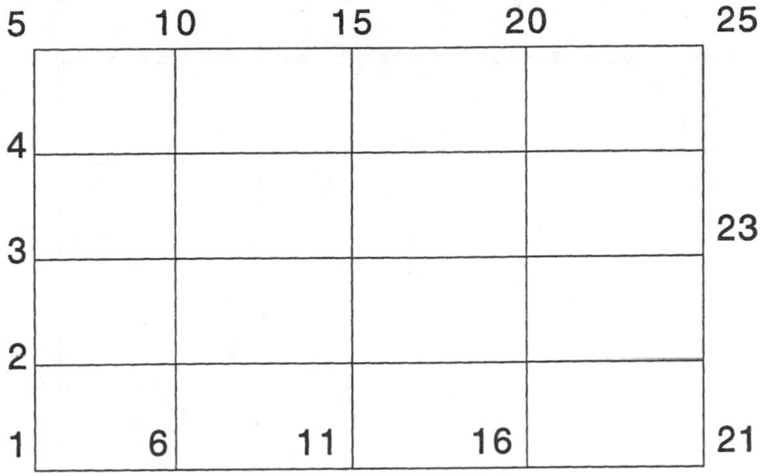

FIGURE 4.8
Finite element model with node numbers.

known heat fluxes. A noise level of 0.5 degrees was added to the measurements, which varied from 0 to 336 degrees.

The advantage of using matrix notation is now realized because the equations for solving the general inverse problem (Section 1.5) remain the same. The state variables, which now represent the temperature vector **T**, are now of length 25 and that of the forcing function **q** is of length 2, representing the two unknown heat fluxes. The exponential matrix method was used to transform the continuous model to a discrete one.

As was the case for the one-dimensional problem, a series of inverse problems were solved varying only the smoothing parameter b. Figure 4.9 shows the cross-validation function V(b) and the true least squares error S(b), which was computed using both heat fluxes. Judging from Figure 4.9, it can be seen that, in this case, the optimal value of b is not estimated very well by the minimum of V(b). If one were to use the solution indicated by minimizing V(b) (b = 8.79E–5), one would obtain the estimates of the heat fluxes shown in Figures 4.10 and 4.11. Although the estimates are very noisy, they still reflect the true nature of the heat fluxes.

One technique for smoothing the unknown heat fluxes is that of first-order regularization (Beck et al., 1985). The idea is to regulate the first derivative of the heat fluxes instead of the heat

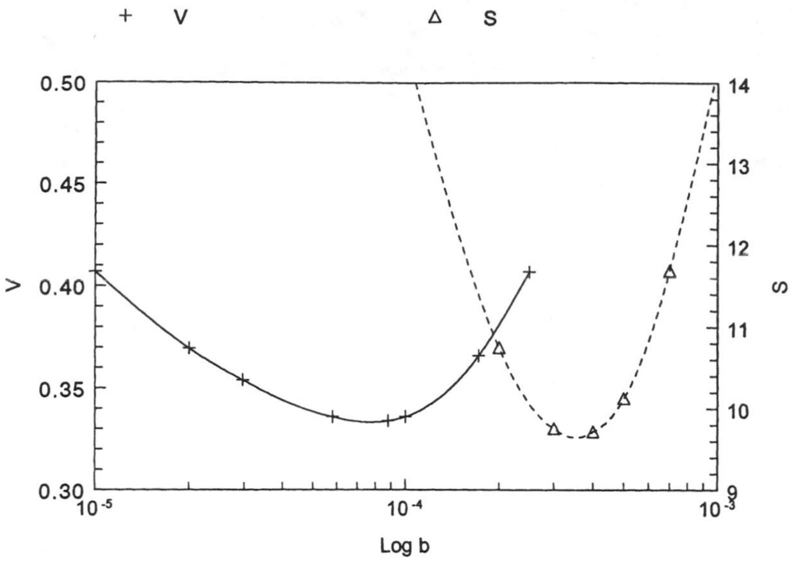

FIGURE 4.9
Variation of S and V with the parameter b two-dimensional example.

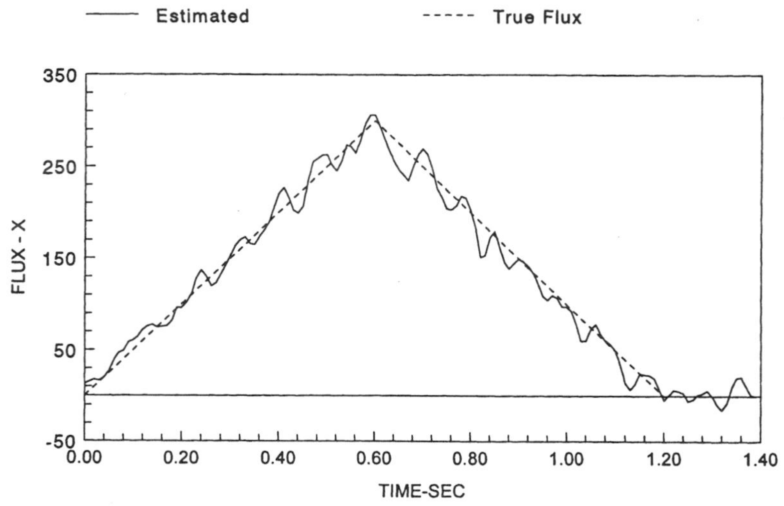

FIGURE 4.10
Left side heat flux (b = 8.79E–5).

fluxes themselves. This is done by adjoining the following differential equations to the thermal model

$$\dot{q}_1 = r_1 \qquad \dot{q}_2 = r_2 \tag{4.3-1}$$

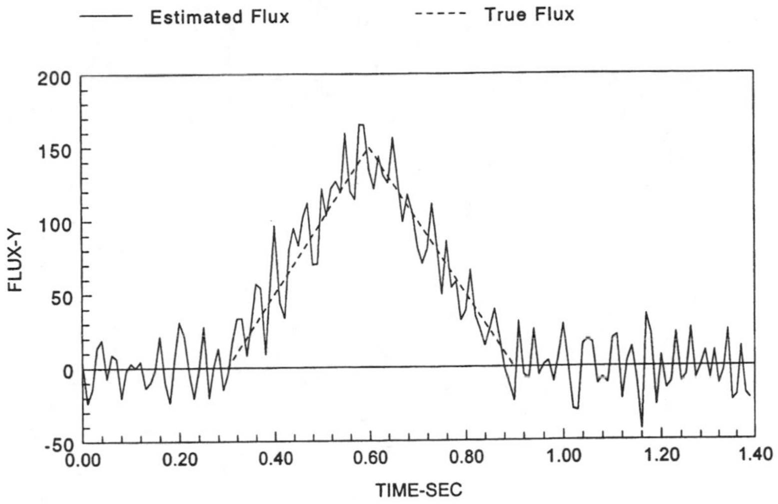

FIGURE 4.11
Top side heat flux (b = 8.79E–5).

The new system becomes

$$\begin{bmatrix} C & 0 \\ 0 & I \end{bmatrix}\begin{bmatrix} \dot{T} \\ \dot{q} \end{bmatrix} = \begin{bmatrix} K & P \\ 0 & 0 \end{bmatrix}\begin{bmatrix} T \\ q \end{bmatrix} + \begin{bmatrix} 0 \\ I \end{bmatrix}[r] \qquad (4.3\text{-}2)$$

and the error expression (Equation 1.5-4) now uses r_1 and r_2 in the regulating term. The variables r_1 and r_2 now represent the unknown functions. Again the matrix formulas remain the same, only the state vector now includes the 25 temperatures and the 2 fluxes.

The results using first-order regularization are shown in Figure 4.12, which shows that V(b) now successfully estimates the optimal regularization parameter b^*. The resulting heat flux estimates are shown in Figures 4.13 and 4.14. They are now very smooth and excellent representations of the true heat fluxes.

Choosing the order of regularization beforehand is difficult. One simple guideline is that if the heat flux estimates still appear "noisy" after selecting the smoothing parameter b by minimizing V(b), then a first-order regularization may yield better results. This is due to the fact that the higher-order regularizations are made on the derivatives of the heat fluxes and will always tend to smooth out the flux estimates.

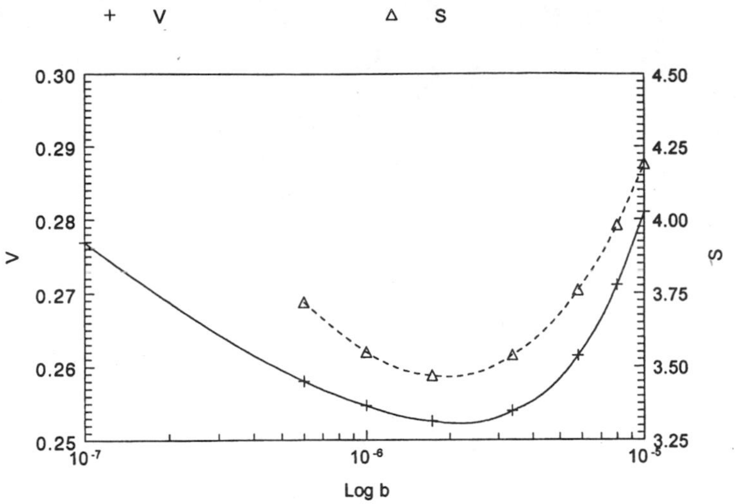

FIGURE 4.12
First-order regularization — variation of s and V with b.

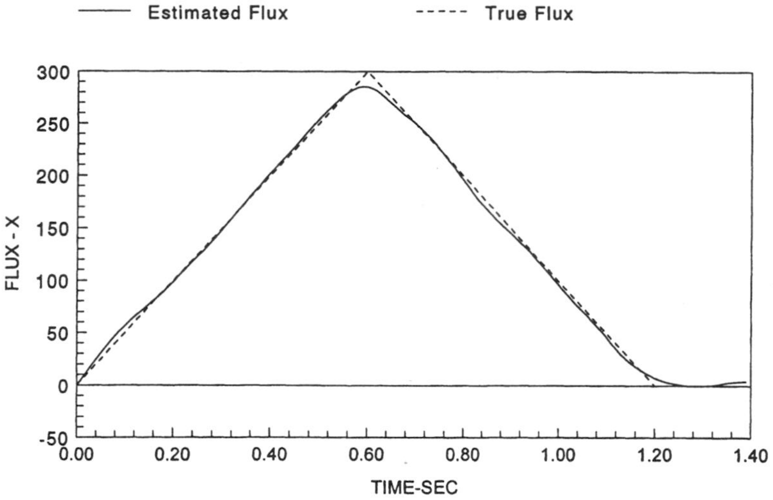

FIGURE 4.13
Left side heat flux — first-order

Exercise

Represent the system equations for a second-order regularization (similar to Equation 4.3-2)

$$q_1 = s_1 \qquad q_2 = s_2$$

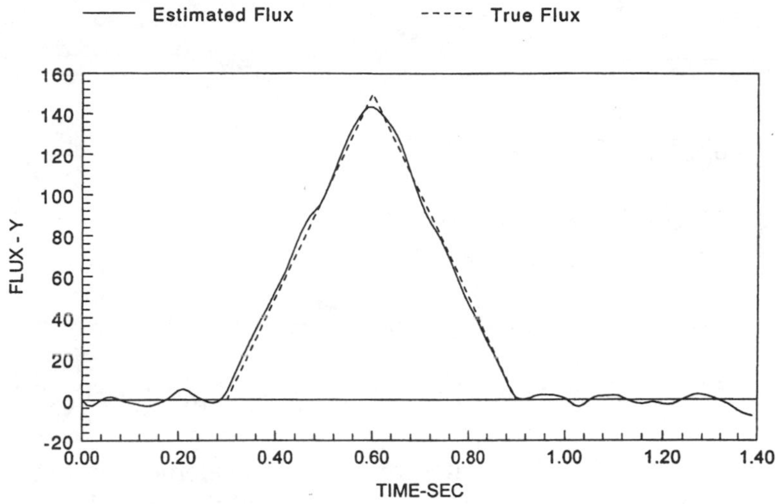

FIGURE 4.14
Top side heat flux — first-order regularization (b = 1.137E–6).

4.3.1 Chebyshev Polynomials

In two-dimensional problems the heat flux can vary over a surface in both time and space. The usual approach is to divide the heat flux into separate zones and assume that it is uniform in each zone. In this manner each heat flux is only a function of time, which can then be directly estimated with the inverse methods.

An alternate approach is to assume that the heat flux distribution over a surface is smooth and can be represented with Chebyshev polynomials (other polynomials could also be used). Thus for any surface of length L let

$$q(x,t) = \sum_{i=1}^{N} a_i(t) C_i(x) \tag{4.3.1-1}$$

where $C_i(x)$ represents the Chebyshev polynomials and $a_i(t)$ represents the unknown functions of time. The variable x represents a local coordinate along any surface. The first five Chebyshev polynomials over the interval [0,L] are (Lanczos, 1964)

$$C_0 = 1$$

$$C_1 = 2(x/L) - 1$$

$$C_2 = 8(x/L)^2 - 8(x/L) + 1$$

$$C_3 = 32(x/L)^3 - 48(x/L)^2 + 18(x/L) - 1$$

$$C_4 = 128(x/L)^4 - 256(x/L)^3 + 160(x/L)^2 - 32(x/L) + 1$$

$$C_5 = 512(x/L)^5 - 1280(x/L)^4 + 1120(x/L)^3 - 400(x/L)^2 + 50(x/L) - 1$$

The heat flux is then integrated over the elements representing the surface, and the integral is equally distributed between the two nodes of the element. In this manner, the participation matrix **P** of Equation 1.5-1 can be constructed. The unknown functions are now the coefficients $a_i(t)$. This participation matrix will now be fully populated since each coefficient a_i affects all the nodes along the surface.

After the coefficients are estimated by the inverse method, the heat flux must be reconstructed using Equation 4.3.1-1. In this manner, a smooth heat flux over the surface is obtained for each timestep. This type of analysis illustrates the versatility of the inverse method in that any special knowledge of a system can be easily incorporated into the general framework of the methodology.

4.4 EIGENVALUE REDUCTION TECHNIQUE

For practical problems the size of the models can become very large, which will result in unreasonable computational effort. Finite element models with 4000 nodes are not uncommon. For large linear models such as these, an eigenvalue reduction technique can be used to reduce the order of the system and yet retain the complexity of the original model. Eigenvalue analysis has been used extensively in the area of structural dynamics (Clough and Penzien, 1975) and has been proposed for heat conduction analysis (Shih and Skladany, 1983). In this section we will outline the application of an eigenvalue technique in conjunction with the inverse method to estimate the unknown heat fluxes (Tandy et al., 1986).

The governing equation for heat conduction in a solid is

$$\mathbf{C}\dot{\mathbf{T}} = \mathbf{KT} + \mathbf{Pq} \tag{4.4-1}$$

The basis for the eigenvalue analysis is to represent the temperature vector **T** as a linear combination of constant vectors φ_i times scalars that vary with time, $u_i(t)$. That is

$$\mathbf{T}(t) = \varphi_1 u_1(t) + \varphi_2 u_2(t) + \mathrm{K} \ + \varphi_N u_N(t) \tag{4.4-2}$$

where N is the length of the temperature vector. For heat conduction systems, the **C** and **K** are positive definite symmetric matrices. In this case the vectors φ_i can be found that satisfy the following relationships (Jennings, 1978):

$$\mathbf{K}\varphi_i = \mathbf{C}\varphi_i w_i \quad \text{for all i} \tag{4.4-3}$$

where w_i are the eigenvalues associated with each vector φ_i. These vectors can also be normalized so that

$$(\mathbf{C}\varphi_i, \varphi_i) = 1.0 \tag{4.4-4}$$

$$(\mathbf{K}\varphi_i, \varphi_i) = w_i \tag{4.4-5}$$

These vectors also satisfy the following orthogonality conditions:

$$(\mathbf{C}\varphi_i, \varphi_j) = 0.0 \tag{4.4-6}$$

$$(\mathbf{K}\varphi_i, \varphi_j) = 0.0 \tag{4.4-7}$$

for all i not equal to j.

As an example, let us assume that only two eigenvectors will be used to approximate **T**, which is of length N much greater than 2.

$$\mathbf{T}(t) \sim \varphi_1 u_1(t) + \varphi_2 u_2(t) \tag{4.4-8}$$

Using this relationship in Equation 4.4-1 gives

$$\mathbf{C}\varphi_1 u_1(t) + \mathbf{C}\varphi_2 u_2(t) = \mathbf{K}\varphi_1 u_1(t) + \mathbf{K}\varphi_2 u_2(t) + \mathbf{Pq} \tag{4.4-9}$$

Taking the inner product of this equation with φ_i and using the orthogonality conditions gives

$$\dot{u}_1 = w_1 u_1 + (\varphi_1, \mathbf{Pq}) \tag{4.4-10}$$

Similarly, by taking the inner product with φ_2 gives an equation for u_2 as

$$\dot{u}_2 = w_2 u_2 + (\varphi_2, \mathbf{Pq}) \tag{4.4-11}$$

These differential equations now represent the model; the unknowns are still **q**; and the measurements **z** are transformed using the approximation of Equation 4.4-8. Thus

$$\mathbf{z}(t) = \mathbf{QT}(t) \sim \mathbf{Q}\varphi_1 u_1(t) + \mathbf{Q}\varphi_2 u_2(t) \tag{4.4-12}$$

In this way the system of size N has been replaced by one of order 2. A methodology for selecting the proper eigenvectors is outlined in Tandy et al. (1986), where several numerical examples are presented.

The restrictions on this type of analysis is that the system must be linear. Also, the computation of the eigenvectors and eigenvalues is not trivial. In the case of the heat conduction equations, several standard methods for eigenvector extraction are available (Jennings, 1978; Press et al., 1990). With more practical experience, this type of analysis will prove to be very useful for large models.

4.5 L-CURVE ANALYSIS

The use of generalized cross validation (GCV) to select the optimum regularization parameter has been previously discussed. GCV worked extremely well but requires the computation of the trace of a global solution matrix that relates all of the measurements to the estimated temperatures. When dealing with finite-element models with thousands of nodes, this method becomes computationally expensive and impractical. For this reason, the L-curve method, recently proposed by Hansen (1992), may prove to be very useful. In this section we will investigate, numerically, the use of the L-curve to select the regularization parameter for a finite-element inverse heat conduction problem. Hansen presented the L-curve method in a general linear algebraic setting where one wishes to minimize the norm of the residual vector adjoined (via a parameter) with a seminorm of the solution. This is known as *Tikhonov regularization*. The seminorm usually represents a numerical approximation to the second derivative of the solution. The L-curve is a plot of the seminorm of the solution versus the residual norm. Hansen points out that the practical use of this plot was first suggested by Lawson and Hanson (1974). With some assumptions, Hansen's analysis shows that the L-curve will depend continuously on the smoothing parameter and that it will always have a corner and that corner estimates the optimal smoothing parameter. This corner is most easily seen in a log-log plot.

In our formulation of the inverse problem the residual norm corresponds to the first term of Equation 3.3-1, which represents the error in matching the data. The seminorm of the derivative of the solution corresponds to the second term of Equation 3.3-1 or alternatively, the derivative of the heat fluxes. In a general formulation of the inverse problem, it is possible to replace the regularization term with a derivative of the heat fluxes in place of the heat fluxes themselves. This is easily accomplished by adjoining the heat fluxes to the state variables and solving for the derivatives of the heat fluxes (see Section 4.3 for more details). Thus, in our application, the following norms will be plotted to produce the L-curves.

$$E_{norm}^2 = \sum_{j=1}^{N} \left(\mathbf{d}_j - \mathbf{U}\mathbf{T}_j, \mathbf{d}_j - \mathbf{U}\mathbf{T}_j\right) \qquad (4.5\text{-}1)$$

$$F_{norm}^2 = \sum_{j=1}^{N} \left(\mathbf{r}_j, \mathbf{r}_j\right) \qquad (4.5\text{-}2)$$

where

$$\mathbf{q}_{j+1} = \mathbf{q}_j + \mathbf{r}_j \tag{4.5-3}$$

The example represents a two-dimensional finite-element model involving two unknown heat fluxes and two temperature measurements. A 0.4 by 0.8 rectangle was modeled with 25 nodes and 16 quadrilateral elements. Unit properties were used. The heat fluxes were applied to two adjacent sides, with the other sides insulated. The measurements were located at the midpoints of the insulated sides. A time increment of 0.01 units was used. A noise level of 0.5 degrees was added to the simulated measurements, which varied from 0 to 336 degrees.

The regularization problem was solved using several values of the smoothing parameter b. The results are shown in Figure 4.15, which plots the F_{norm} versus the E_{norm} on log-log scales. The

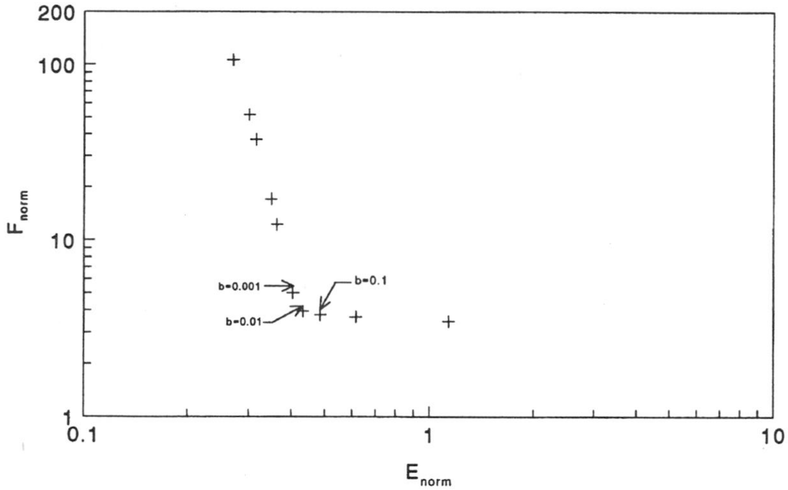

FIGURE 4.15
L-curve, F_{norm} versus E_{norm}.

L-shape characteristic of the curve is indeed present. The values of the smoothing parameter are shown for the data in the proximity of the corner. It only remains to validate that the optimal value of the smoothing curve does occur at the corner.

One advantage of testing methods and theories with simulated data is that one knows the answer beforehand. Figure 4.16

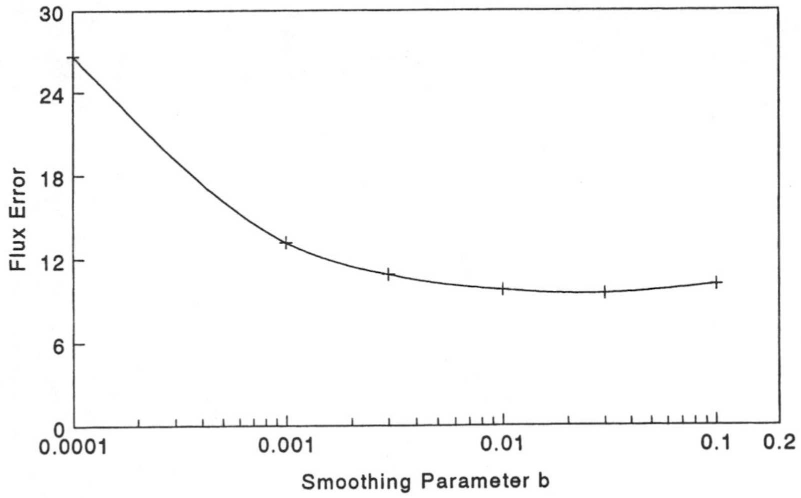

FIGURE 4.16
True flux error versus b.

shows a plot of the least squares error between the true heat fluxes and the estimated ones for various smoothing parameters. From this figure, the optimal smoothing parameter would be chosen as 0.03. From Figure 4.15, the corner of the L-curve is judged to be at 0.01. The estimated heat fluxes are not extremely sensitive to the parameter and either of these values would yield excellent estimates of the heat fluxes. The estimated and true heat fluxes are shown in Figures 4.17 and 4.18 for the smoothing parameter of 0.01.

This next example is taken from Trujillo and Wallis (1989) and represents real experimental data (see Section 7.3.2). A quenching experiment was carried out with an instrumented disk, (10.5 inches in diameter by 2.75 inches thick) made from Alloy 718. The disk was heated to 2150°F in a gas-fired furnace before being transferred to an oil tank where it was quenched. An axisymmetric finite element model consisting of 400 nodes ($n = 400$) was used to represent the disk. Ten thermocouples ($m_g = 10$) were placed in the disk to capture the transient data. The model was then used to estimate seven heat flux histories ($m = 7$) around the perimeter of the disk. The thermocouples were sampled every two seconds for a total of 180 points ($N = 180$). Also, this problem is nonlinear because of the temperature dependence of the thermodynamic properties of the metal.

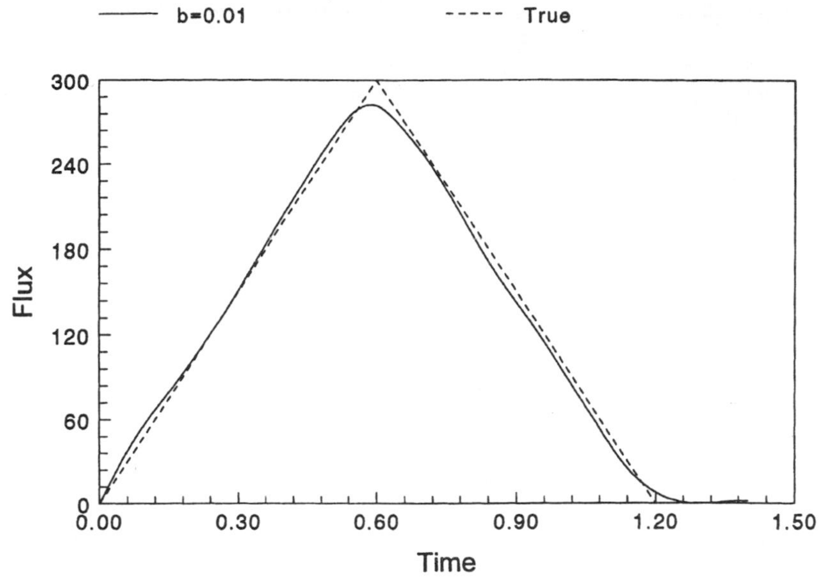

FIGURE 4.17
Estimated and true flux 1.

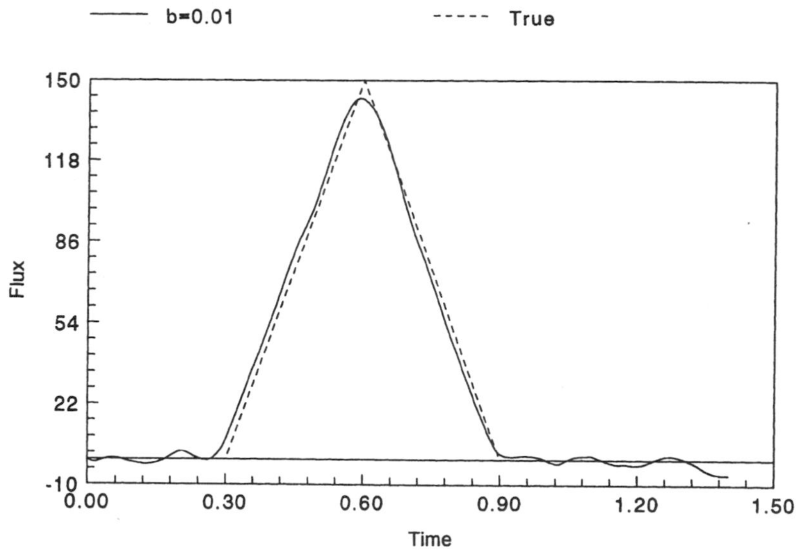

FIGURE 4.18
Estimated and true flux 2.

The L-curve was constructed by solving the regularization problem using several values of the smoothing parameter b. The results are shown in Figure 4.19 which plots the F_{norm} versus the

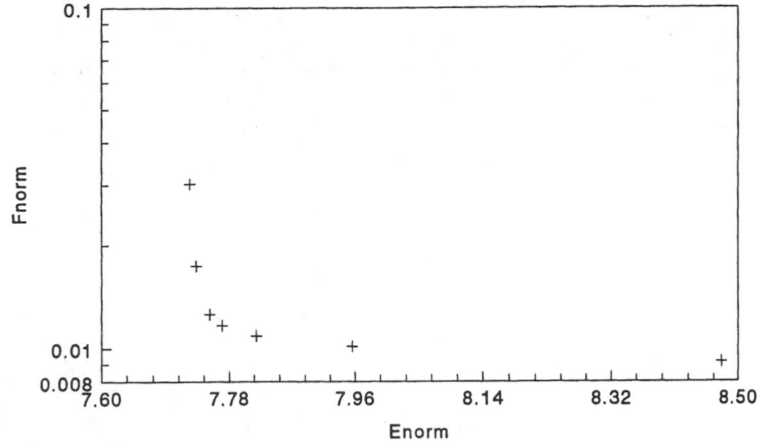

FIGURE 4.19
L-curve, experimental data.

E_{norm} on a log-linear scale. For this case the E_{norm} did not vary much in magnitude. However, the L-shape characteristic of the curve is indeed present. Since this case represents real data there is no "true" answer to help evaluate the performance of the L-curve, but the corner value corresponds very well to the value of the smoothing parameter that was chosen by Trujillo and Wallis (1989) based on experience and intuition.

The results of the above example give one confidence that the L-curve is indeed a good method for selecting the optimal smoothing parameter. As Hansen indicates, the L-curve and generalized cross validation would give the same parameter for most cases with white noise and even with filtered white noise. Another advantage of the L-curve method is that it can be plotted with readily available information, while constructing the GCV curve requires some additional computations. Thus, in computer programs that solve very large models the L-curve method is the only available practical method. It is also extremely encouraging that the L-curve gives excellent results for experimental data.

Chapter 5

The Inverse Structural Dynamics Problem

5.1 INTRODUCTION

In static problems one investigates the behavior of structures under loads that are imposed slowly and therefore considered static. In reality, no structural loads are really static since they need to be applied in some manner to the structure and this requires a time variation in the forces. In most cases the dynamic response of a structure will be complex. This complexity can be related to the number of degrees of freedom of the structure. In this case, the degrees of freedom are identical to the number of independent coordinates that are needed to specify the configuration or position of the system. A continuous structure has, for example, an infinite number of degrees of freedom, whereas a rigid body moving in a Cartesian coordinate system has six degrees of freedom.

If the time interval of the load is completely known together with the system parameters, then the analysis of the response of the structural system to the known load is defined as being deterministic and the problem can be solved directly. In solving the direct problem several conditions must be satisfied. For example, in solving a direct beam problem, the governing equations, boundary conditions, initial conditions, and the forces acting on the beam must be known. When one or more of these quantities are lacking or incomplete, a direct problem cannot be performed and it becomes a type of inverse problem. In this section, however, an inverse problem will be defined as one in which the forcing term is the unknown. For example, if the time variation of the load is unknown but the response of the system can be measured (i.e., displacement, velocity, acceleration, stress or strain versus time) then we have an application of the inverse structural dynamics problem.

There has been only a limited amount of research performed in the area of inverse structural dynamics problems. Some work that has been done includes Busby and Trujillo (1987), Hollandsworth and Busby (1989), Simonian (1981), and Trujillo (1978). In the literature many authors use "inverse problem" synonymously with the term "identification problem." In our terminology, identification problems relate to determining an unknown system parameter such as mass, stiffness, damping coefficient, etc. using the known input and output response of the system. This type of inverse problem will be covered in Chapter 8.

In the present chapter dynamic programming and cross validation will be applied to various inverse structural dynamics problems ranging from a simple spring mass system to a structure with many degrees of freedom.

The inverse technique, described in Chapter 3, consists of a backward time sweeping phase followed by a standard forward time sweeping phase. The backward sweep calculates and stores matrix and vector recursive values for a total of N time steps. These recursive relations are dependent upon the measurements (displacement, velocity, acceleration, stress or strain) and the system parameters of a mathematical model. The forward process utilizes these relations so that a predicated measurement and force response is calculated for each time step. The analysis presented in this chapter deals with the predication of a single aperiodic load using one or more measurements but is easily extended to several loads. In addition, the method allows the use of more than one type of measurement at the same time, such as velocity and strain versus time data.

5.2 SINGLE-DEGREE-OF-FREEDOM

When a system is constrained so that it can only have motion in one direction, it is defined as a single or a one-degree-of-freedom system. Thus, only one geometric quantity is required to completely describe its movement or motion. This geometric quantity could be a displacement, a voltage, a current, or an angle. A system that requires more than one geometric quantity

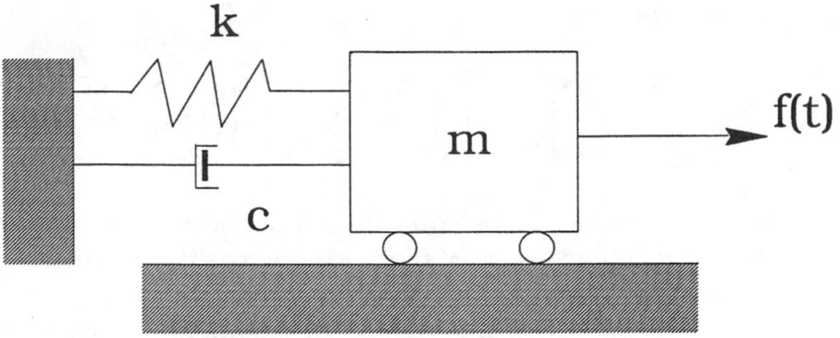

FIGURE 5.1
Single-degree-of-freedom system.

for its complete description is called a multi-degree-of-freedom system.

Consider the single-degree-of-freedom system shown in Figure 5.1. The system has a mass m, rolls on a smooth surface, and is attached to a spring and dashpot having stiffness k and a coefficient of viscous damping c. This viscous-damped oscillator is subjected to an impulse force f(t). The differential equation of motion is given as

$$m\ddot{x} + c\dot{x} + kx = f(t) \tag{5.2-1}$$

Thus Equation 5.2.1 is a typical nonhomogeneous, linear differential equation of the second order with constant coefficients, whose direct solution can easily be obtained by using the methods of Chapter 2. The initial conditions are given by

$$x(0) = x_0, \qquad \dot{x}(0) = v_0$$

Equation 5.2.1 can be rewritten as two first-order differential equations by defining

$$x_1(t) = x(t), \qquad x_2(t) = \dot{x}(t)$$

This leads to the following set of equations

$$\begin{aligned} \dot{x}_1 &= x_2 \\ m\dot{x}_2 &= -cx_2 - kx_1 + f(t) \end{aligned} \tag{5.2-2}$$

or in matrix form as

$$\begin{bmatrix} 1 & 0 \\ 0 & m \end{bmatrix} \begin{Bmatrix} \dot{x}_1 \\ \dot{x}_2 \end{Bmatrix} = \begin{bmatrix} 0 & 1 \\ -k & -c \end{bmatrix} \begin{Bmatrix} x_1 \\ x_2 \end{Bmatrix} + \begin{Bmatrix} 0 \\ 1 \end{Bmatrix} f(t) \qquad (5.2\text{-}3)$$

As shown in Chapter 2, the solution to the above equation can be obtained using a variety of methods such as the exponential technique, backward implicit, or Newmark beta method.

As an illustration of an inverse problem, the above one-degree-of-freedom system was first subjected to the forcing functions shown in Figure 5.2: (a) half triangular load, (b) square load, and (c) a full triangular load. The values used for m, k, and c were

$$m = 100 \text{ lb.}, \quad k = 1000 \text{ lb/in}, \quad c = 0.0$$

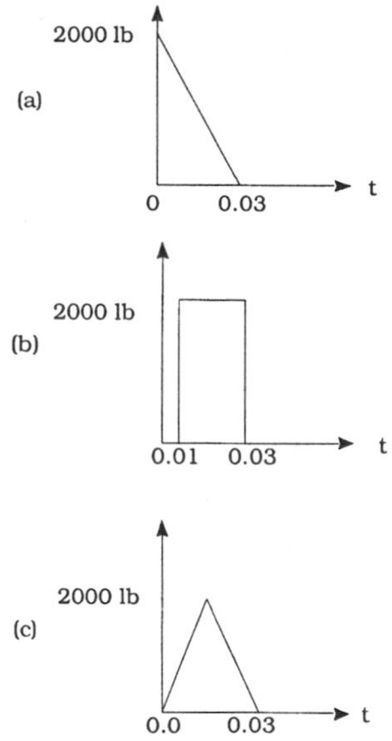

FIGURE 5.2
Forcing function.

A direct integration technique was used to generate simulated velocity measurements. Noise was then added to simulate real measurements. A sampling time of 0.001 was used for all cases. The noisy velocities (using a σ of 5.0) for cases (a – c) are shown in Figures 5.3 through 5.5. These measurements were then used

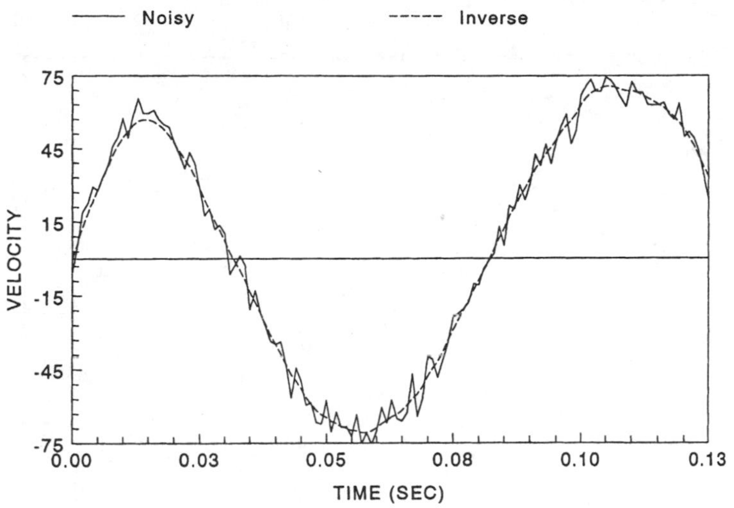

FIGURE 5.3
Velocity versus time — half triangular load.

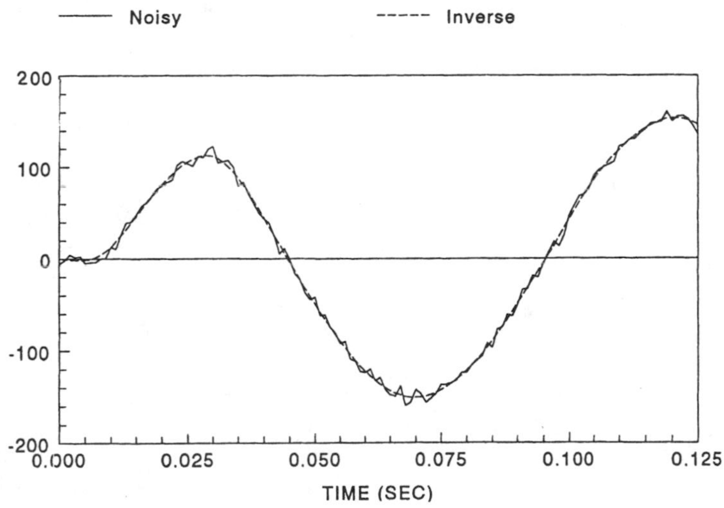

FIGURE 5.4
Velocity versus time — square load.

in an inverse solution using the techniques of Chapter 2. The above figures also show how the inverse solution smooths the noisy data. The forces predicted by the inverse technique for cases (a – c) are shown in Figures 5.6 through 5.8. It is noted that the force estimates are still a little noisy, however, they still match fairly well. The optimal smoothing parameters were determined using generalized cross validation.

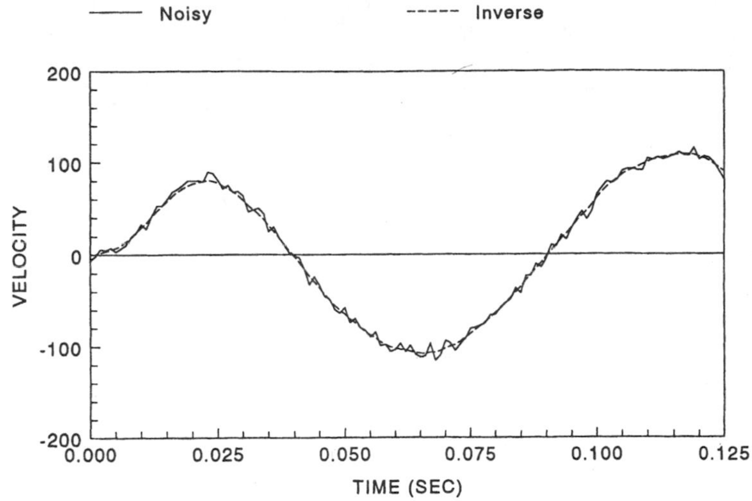

FIGURE 5.5
Velocity versus time — full triangular load.

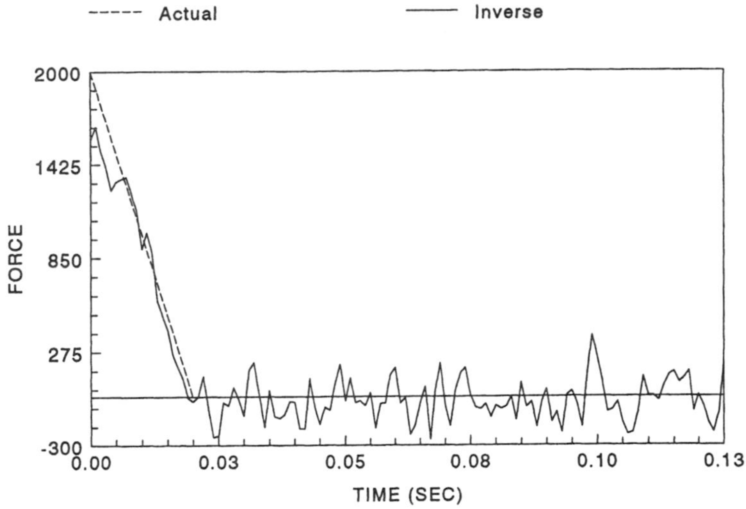

FIGURE 5.6
Half triangular load.

To further smooth the unknown forces, the technique of first-order regularization can be used as it was in the heat transfer problem. Thus Equation 5.2.3 can be rewritten as

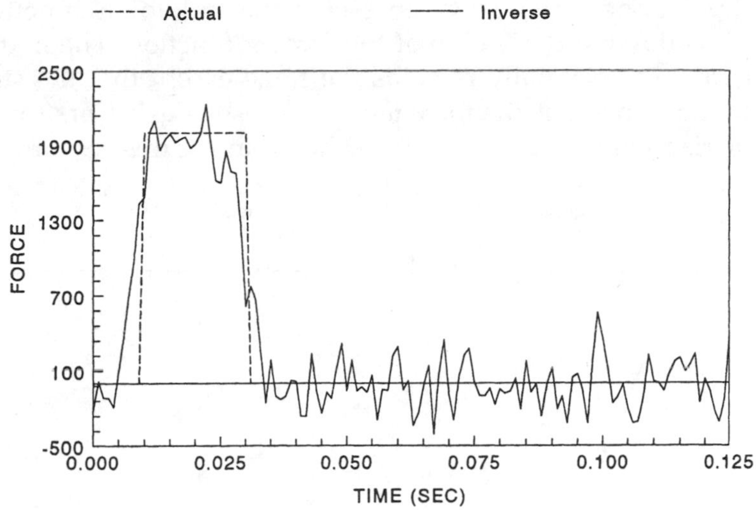

FIGURE 5.7
Square load.

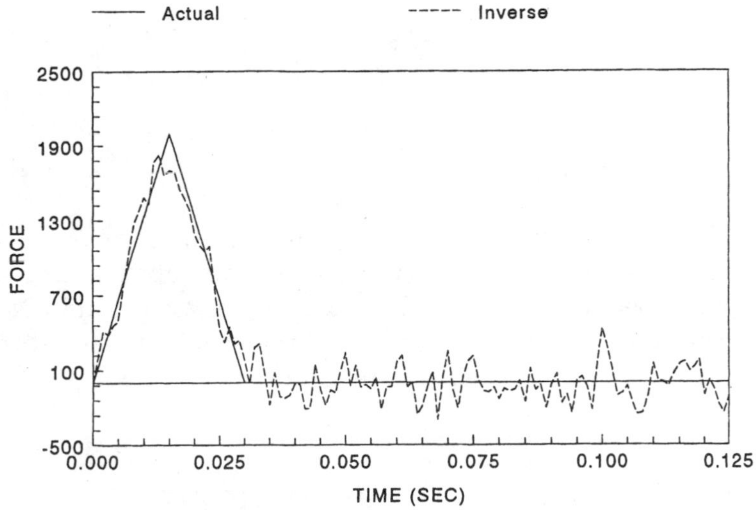

FIGURE 5.8
Full triangular load.

$$\begin{bmatrix} 1 & 0 & 0 \\ 0 & m & 0 \\ 0 & 0 & 1 \end{bmatrix} \begin{Bmatrix} \dot{x}_1 \\ \dot{x}_2 \\ \dot{f} \end{Bmatrix} = \begin{bmatrix} 0 & 1 & 0 \\ -k & -c & 1 \\ 0 & 0 & 0 \end{bmatrix} \begin{Bmatrix} x_1 \\ x_2 \\ f \end{Bmatrix} + \begin{Bmatrix} 0 \\ 0 \\ 1 \end{Bmatrix} r(t) \tag{5.2-4}$$

The variable r(t) now represents the unknown function, which is the first derivative of the forcing function. The matrix that we had essentially remains the same except that the state vector now includes the force term. The results using first-order regularization for cases (a – c) are shown in Figures 5.9 through

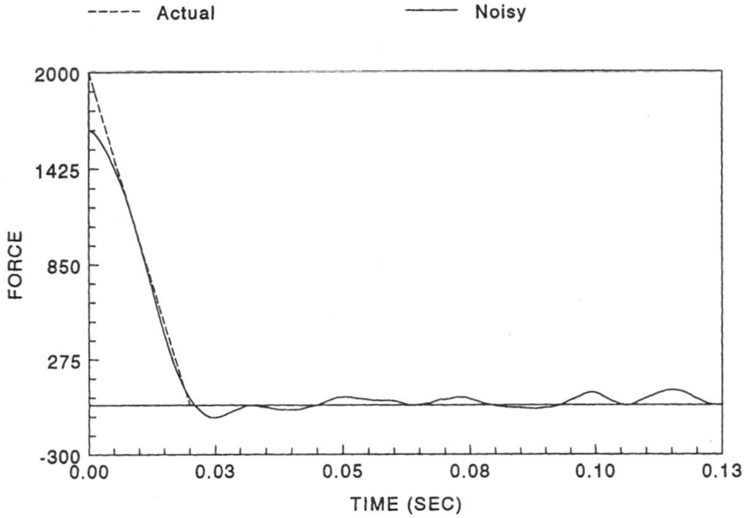

FIGURE 5.9
Half triangular load — first-order regularization.

5.11. The forces even with $\sigma = 5.0$ are smooth and show excellent representations to the true external forces. The b optimal smoothing parameter values obtained for these cases are given in Table 5.1. Experience indicates that for dynamics problems first-order regularization is almost always required.

5.3 CANTILEVER BEAM PROBLEM

A beam will be used to demonstrate the application to a one-dimensional structural dynamics problem. The inverse dynamics problem has been previously investigated by Busby and Trujillo (1987), who showed how to reduce the order of a system with the use of eigenvector analysis. Lim and Pilkey (1992) investigated a large truss structure subjected to an unknown dynamic force. They showed how to select the important modes and how

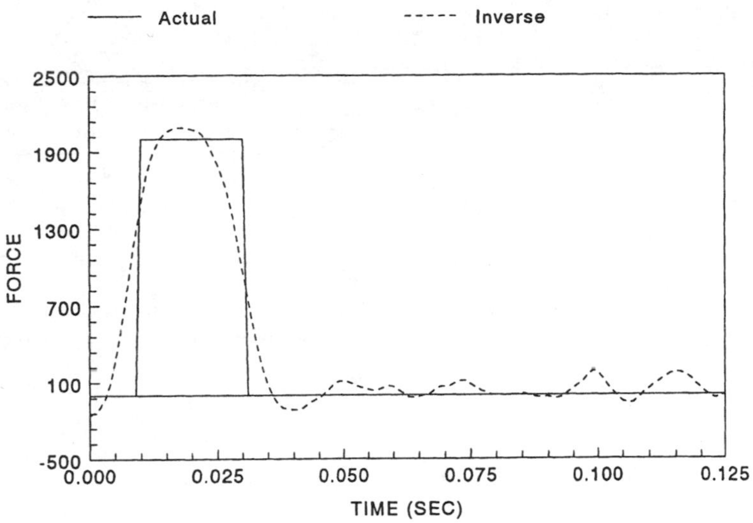

FIGURE 5.10
First-order regularization.

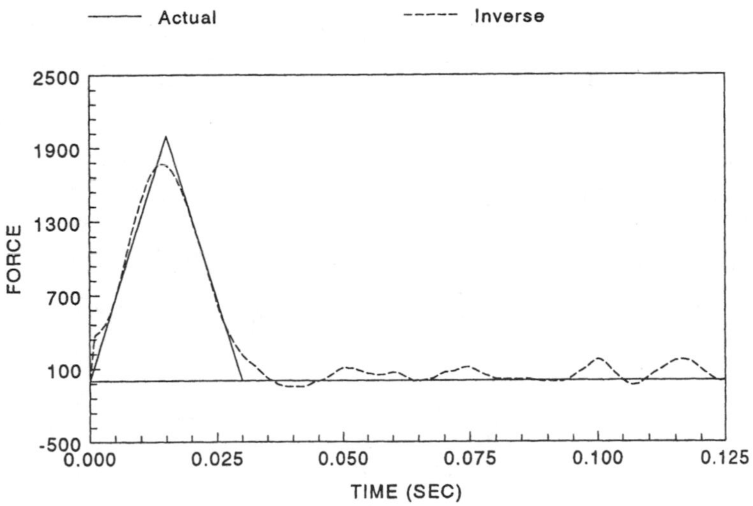

FIGURE 5.11
Full triangular load — first-order regularization.

to determine the best measurement locations. Their method was restrictive in that it required the simultaneous measurement of displacement, velocity, and acceleration at each measurement location. D'Cruz, Crisp, and Ryall (1992) solved a slightly different inverse problem in that the applied force was known to be

TABLE 5.1

b optimal smoothing parameter values

Force Loading	Smoothing Parameter (b)	Smoothing Parameter First Order Regularization
(a) Half Triangular	1.18070E–04	4.76450E–03
(b) Square	6.91826 E–05	1.19629E–03
(c) Triangular	9.96165E–05	1.54645E–03

a harmonic but its location was unknown. In our investigation the location of the force is known but the time history is unknown.

5.3.1 Mathematical Model

A cantilever beam will be converted to a vector-matrix differential equation using a finite element formulation.

The differential equation for a beam is given as

$$\frac{\partial^2 u}{\partial t^2} + \left(\frac{EI}{\rho A}\right)\left(\frac{\partial^4 u}{\partial x^4}\right) = 0 \qquad (5.3.1\text{-}1)$$

where u is the transverse displacement, E is Young's modulus, I is the moment of inertia, ρ is the density, and A is the cross-sectional area.

The finite element approximation for a beam element of length L is given as

$$k = \left(\frac{EI}{L^3}\right)\begin{bmatrix} 12 & -6L & -12 & -6L \\ -6L & 4L^2 & 6L & 2L^2 \\ -12 & 6L & 12 & 6L \\ -6L & 2L^2 & 6L & 4L^2 \end{bmatrix} \qquad (5.3.1\text{-}2)$$

for the stiffness matrix and

$$m = \left(\frac{\rho AL}{420}\right)\begin{bmatrix} 156 & -22L & 54 & 13L \\ -22L & 4L^2 & -13L & -3L^2 \\ 54 & -13L & 156 & 22L \\ 13L & -3L^2 & 22L & 4L^2 \end{bmatrix} \qquad (5.3.1\text{-}3)$$

for the mass matrix. The degrees of freedom for these elements are the transverse displacement and the rotation at each end of the beam [u_1 θ_1 u_2 θ_2]. The strain at any location x along the beam element is given in terms of these degrees of freedom by

$$\varepsilon_x = \left(\frac{-y}{L^3}\right)\Big[(12x - 6L)u_1 - 12(x - 6L)\theta_1 + L(6x - 4L)u_2 + L(6x - 2L)\theta_2\Big] \tag{5.3.1-4}$$

where y represents the distance from the neutral axis of the beam.

A very common approximation is to condense out the rotational degrees of freedom to reduce the order of the system. This condensation relates all of the rotational degrees of freedom to the transverse ones. This approximation will, in turn, also affect the calculation of the strains which are now related to all of the transverse displacements. The final dynamic system is represented as

$$\mathbf{M}\ddot{\mathbf{U}} + \mathbf{K}\mathbf{U} = \mathbf{F}(t) \tag{5.3.1-5}$$

where $\mathbf{U}$ is a vector containing all the transverse displacements of the model, $\mathbf{M}$ is the assembled mass matrix, $\mathbf{K}$ is the assembled stiffness matrix, and $\mathbf{F}(t)$ represents the applied forces.

Equation 5.3.1-5 can be converted to a first-order discrete system

$$\mathbf{x}_{j+1} = \mathbf{T}\mathbf{x}_j + \mathbf{P}\mathbf{f}_j \tag{5.3.1-6}$$

where $\mathbf{x}$ represents a vector of length 2n containing the displacements and velocities of the nodes.

Now suppose that a series of measurements have been taken and are represented by the vectors $\mathbf{d}_j$, where the length of $\mathbf{d}_j$ is m. The measurements could represent displacements, velocities, or accelerations. The number of measurements m is usually much less than the number of variables n but greater than the number of known forces. These measurements are related to the state variables $\mathbf{x}_j$ by

$$\mathbf{d}_j \propto \mathbf{Q}\mathbf{x}_j \tag{5.3.1-7}$$

where **Q** is an $(m \times n)$ matrix which associates the measurements to the state variables.

In this example, the measurement will represent the strain in the middle of the first element from the built-in end. The beam is divided into 10 elements. Because of the elimination of the rotational degrees of freedom, this strain is related to all of the transverse displacements, and Q becomes a (1×10) matrix.

The problem is to find the unknown forces $\mathbf{f}_j$ that when used in Equation 5.3.1-6 will force the model to best match the measurements represented by Equation 5.3.1-7. It quickly becomes obvious that an exact match will not work. This is due to the fact that all measurements have some degree of noise, while the models, on the other hand, have usually assumed all kinds of derivatives and smoothness. The data is adjoined to the model with the use of least squares. In vector form, this is represented with a vector inner product (.,.) and would be represented by an error sum over all the data points N.

$$E = \sum_{j=1}^{N} \left(\mathbf{d}_j - \mathbf{Q}\mathbf{x}_j, \mathbf{d}_j - \mathbf{Q}\mathbf{x}_j \right)$$

Even this least squares criterion is not sufficient because a mathematical solution that will minimize E will usually end up with the model exactly matching the data — a situation that is to be avoided. This is where the regularization method enters. By adding a term to the above least squares error

$$E = \sum_{j=1}^{N} \left(\mathbf{d}_j - \mathbf{Q}\mathbf{x}_j, \mathbf{d}_j - \mathbf{Q}\mathbf{x}_j \right) + b\left(\mathbf{f}_j, \mathbf{f}_j \right) \qquad (5.3.1\text{-}8)$$

one can control the amount of smoothness that occurs in the solution by varying the parameter b. This is Tikhonov's method as described in Chapter 3. What is now required of the solution is to best match the data (the first term of Equation 5.3.1-8) but to have some degree of smoothness (the second term of Equation 5.3.1-8).

Two methods used to determine the optimal value of b are the L-curve method and generalized cross validation.

L-Curve Method

As we discussed in Section 4.5, Hansen (1992) presents the L-curve method in a general linear algebraic setting where one wishes to minimize the norm of the residual vector adjoined (via a parameter) with a seminorm of the solution. This is known as Tikhonov regularization. The seminorm usually represents a numerical approximation to the second derivative of the solution. The L-curve is a plot of the seminorm of the solution versus the residual norm for various parameters. Hansen points out that the practical use of this plot was first suggested by Lawson and Hanson (1974). With some assumptions, Hansen's analysis shows that the L-curve will depend continuously on the smoothing parameter and that it will always have a corner and that corner estimates the optimal smoothing parameter. This corner is most easily seen in a log-log plot.

In our formulation of the inverse problem, the residual norm corresponds to the first term of Equation 5.3.1-8, which represents the error in matching the data. The seminorm of the derivative of the solution corresponds to the second term of Equation 5.3.1-8.

Zeroth and First-Order Regularization

The above formulation of the inverse problem is also known as the *zeroth order regularization*. In the general formulation of the inverse problem, it is possible to replace the regularization term with a derivative of the unknown forces in place of the forces themselves. This is easily accomplished by adjoining the forces to the state variables and solving for the derivatives of the forces, $\mathbf{r}_j$. Thus let the forces be represented by

$$\mathbf{f}_{j+1} = \mathbf{f}_j + \mathbf{r}_j \tag{5.3.1-9}$$

The new system is now

$$\begin{Bmatrix} \mathbf{x}_{j+1} \\ \mathbf{f}_{j+1} \end{Bmatrix} = \begin{bmatrix} \mathbf{T} & \mathbf{P} \\ \mathbf{0} & \mathbf{I} \end{bmatrix} \begin{Bmatrix} \mathbf{x}_j \\ \mathbf{f}_j \end{Bmatrix} + \begin{Bmatrix} \mathbf{0} \\ \mathbf{r}_j \end{Bmatrix} \tag{5.3.1-10}$$

and the error formula becomes

$$E = \sum_{j=1}^{N} \left(\mathbf{d}_j - \mathbf{Q}\mathbf{x}_j, \mathbf{d}_j - \mathbf{Q}\mathbf{x}_j\right) + b\left(\mathbf{r}_j, \mathbf{r}_j\right) \qquad (5.3.1\text{-}11)$$

This is known as first-order regularization. In our experience, the first-order regularization works better for structural dynamics problems, especially for relatively large noise levels in the measurements. Thus, in our application, the following norms will be plotted to produce the L-curves.

$$E_{norm}^2 = \sum_{j=1}^{N} \left(\mathbf{d}_j - \mathbf{U}\mathbf{x}_j, \mathbf{d}_j - \mathbf{U}\mathbf{x}_j\right) \qquad (5.3.1\text{-}12)$$

and for zeroth-order regularization

$$F_{norm}^2 = \sum_{j=1}^{N} \left(\mathbf{f}_j, \mathbf{f}_j\right) \qquad (5.3.1\text{-}13)$$

and for the first-order regularization

$$F_{norm}^2 = \sum_{j=1}^{N} \left(\mathbf{r}_j, \mathbf{r}_j\right) \qquad (5.3.1\text{-}14)$$

5.3.2 Example

This example represents a finite-element model of a cantilever beam, Figure 5.12. The beam will be represented with 10 elements with the first node fixed to represent the cantilever. Thus, $n = 10$, but in practical problems n can easily reach 1000. Based on our equations given above for this problem, **T** is a matrix which represents the dynamics of the model, **f** is a vector of length n_g representing the unknown applied forces, and **P** is matrix $(2n \times n_g)$ relating the forces to the system. In this example, there is only one unknown force so that n_g is equal to one. A timestep h represents the difference between the state variables states $\mathbf{x}_j$ and $\mathbf{x}_{j+1}$. The beam analyzed was one meter in length. The cross-section is a rectangle 64 mm × 13 mm. The moment of inertia is 11,717 mm^4 and the Young's modulus E is 72 GPa. The density was 2800 Kg/m^3. This beam was divided into 10 equal elements and the unknown force was applied at the tip. A timestep of 5.0E–4 seconds was used in the simulation. The theoretical first natural frequency of this beam is 10.7 Hz. The first natural frequency of the finite element model was found to be 10.6 Hz, which validates the accuracy of the model.

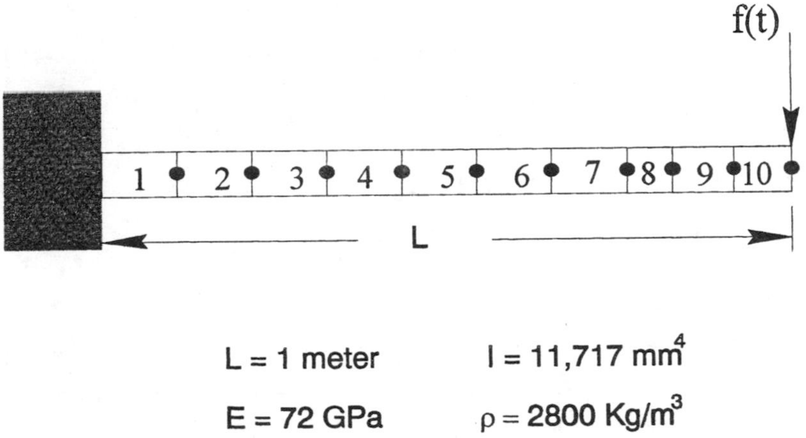

FIGURE 5.12
Cantilever beam finite element model.

A direct solution of this model was used to solve for the displacements and the strain histories of the first element near the built-in end. A Gaussian noise level of 3.0 percent of the maximum strain was added to the simulated measurements.

The zeroth-order regularization problem was solved using several values of the smoothing parameter b. The results are shown in Figure 5.13, which plots the F_{norm} versus the E_{norm} on log-log scales and in Figure 5.14, which shows the GCV function. The L-shape characteristic of the curve is indeed present. The values of the smoothing parameter are shown for the data in the proximity of the corner. The minimum of the GCV curve occurs at b = 1.0E–17. It only remains to validate that the optimal value of the smoothing curve does occur at the corner.

One advantage of testing methods and theories with simulated data is that one knows the answer beforehand. Figure 5.15 shows a plot of the least squares S(b) error between the true force and the estimated one for various smoothing parameters. From

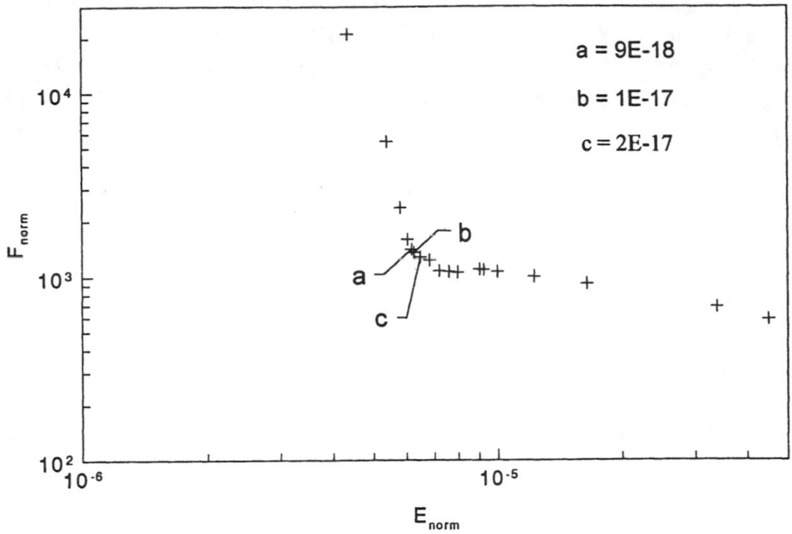

FIGURE 5.13
L-Curve — F_{norm} versus E_{norm}

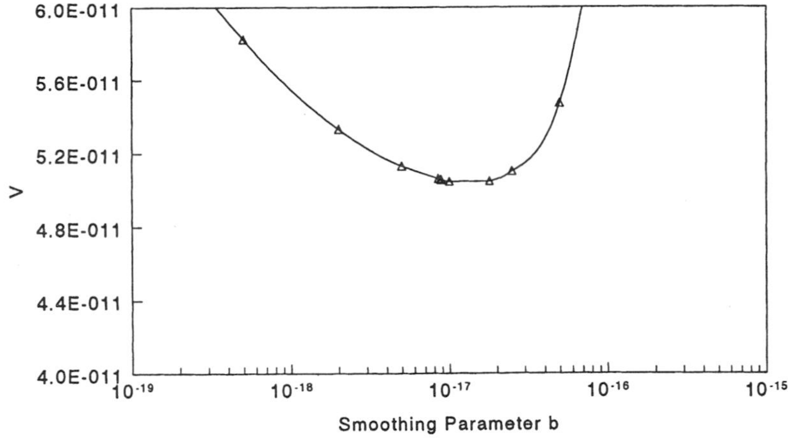

FIGURE 5.14
Generalized cross-validation function.

this figure, the optimal smoothing parameter would be chosen as 1.15E–17. From Figure 5.13, the corner of the L-curve is judged to be at 9.0E–18. The estimated force is not extremely sensitive to the parameter and either of these values would yield excellent estimates. The estimated and true force is shown in Figure 5.16 for a smoothing parameter of 1.0E–17.

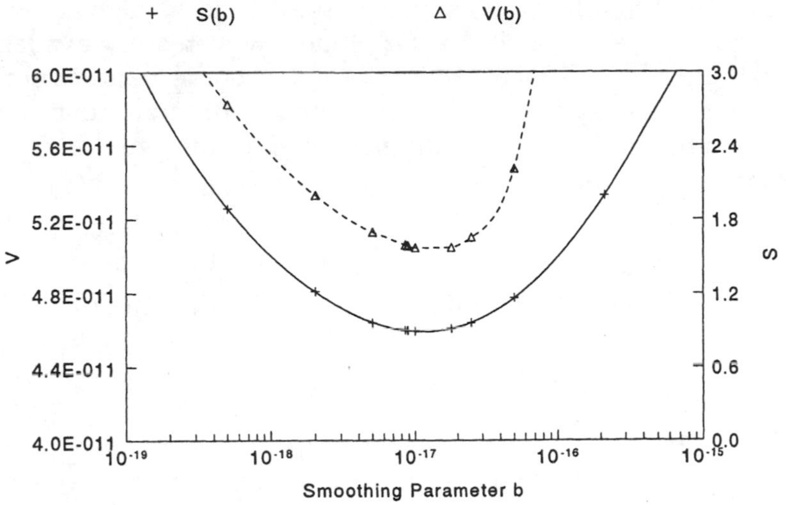

FIGURE 5.15
Variation of S and generalized cross-validation function V with the parameter b.

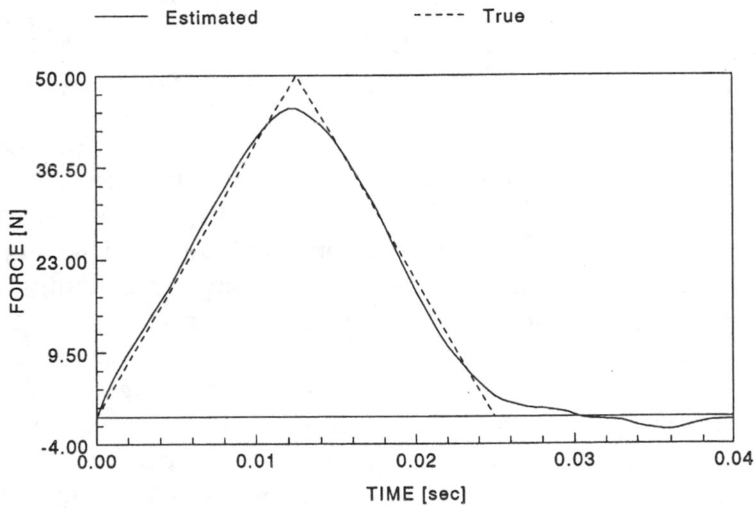

FIGURE 5.16
True and estimated force for b = 1.0E–17.

The results of the above example gives one confidence that generalized cross-validation and the L-curve are indeed good methods for selecting the optimal smoothing parameter. As Hansen indicates, the L-curve and generalized cross validation would give the same parameter for most cases with white noise and even with filtered white noise. Another advantage of the L-curve method is that it can be plotted with readily available information, while constructing the GCV curve requires some additional computations. Thus, in computer programs that solve very large models, the L-curve method is the only available practical method.

5.4 TWO-DIMENSIONAL PLATE PROBLEM

5.4.1 Introduction

Michaels and Pao (1985) investigated the inverse source problem for an infinite plate. They presented an iterative method of deconvolution which was used to determine the orientation and time-dependent amplitude of the force from the transient response of the plate surface at a minimum of two locations given the source location. Lin and Datseris (1988) employed the theory of plates to solve an unknown inverse plate problem with application to position and force sensing. Manoach, Karagiozova, and Hadjikov (1991) solved an inverse problem of an irregularly heated thin circular plate subjected to a pulse loading. They considered two cases: (i) when a restriction on the maximum center deflection is given, and (ii) when the displacement of the plate center as a known function in time is required. D'Cruz, Crisp and Ryall (1992) present a method for determining the location and magnitude of a static point force acting on a simply supported elastic rectangular plate from a number of displacement readings at discrete points on the plate. D'Cruz, Crisp, and Ryall (1992) also solved a slightly different inverse problem in that the applied force was known to be a harmonic, but its location was unknown. In our investigation the location of the force is known, but the time history is unknown.

5.4.2 Mathematical Model

Although the mathematical model can be very general and represent any linear dynamic system, a simply supported square orthotropic plate with first-order shear deformation will be used to investigate the performance of the methods. The simply supported plate will be converted to a vector-matrix differential equation using a finite element formulation (Reddy, 1984).

The differential equation for a plate with first-order shear deformation is given by

$$
\begin{aligned}
&L_1(w,\phi_x,\phi_y)+q=m_1\,\partial^2 w/\partial t^2\\
&L_2(w,\phi_x,\phi_y)=m_2\,\partial^2\phi_x/\partial t^2\\
&L_3(w,\phi_x,\phi_y)=m_2\,\partial^2\phi_y/\partial t^2
\end{aligned}
\tag{5.4.2-1}
$$

where L_1, L_2, and L_3 are linear differential operators defined by

$$
L_1=A_{55}\frac{\partial}{\partial x}\left(\phi_x+\frac{\partial \omega}{\partial x}\right)+A_{44}\frac{\partial}{\partial y}\left(\phi_y+\frac{\partial w}{\partial y}\right)
\tag{5.4.2-2}
$$

$$
\begin{aligned}
L_2=&\frac{\partial}{\partial x}\left(D_{11}\frac{\partial\phi_x}{\partial x}+D_{11}\frac{\partial\phi_y}{\partial y}\right)\\
&+D_{66}\frac{\partial}{\partial y}\left(\frac{\partial\phi_x}{\partial y}+\frac{\partial\phi_y}{\partial x}\right)-A_{55}\left(\phi_x+\frac{\partial w}{\partial x}\right)
\end{aligned}
\tag{5.4.2-3}
$$

$$
\begin{aligned}
L_3=&D_{66}\frac{\partial}{\partial x}\left(\frac{\partial\phi_x}{\partial y}+\frac{\partial\phi_y}{\partial x}\right)\\
&+\frac{\partial}{\partial y}\left(D_{12}\frac{\partial\phi_x}{\partial x}+D_{22}\frac{\partial\phi_y}{\partial y}\right)-A_{45}\left(\phi_y+\frac{\partial w}{\partial y}\right)
\end{aligned}
\tag{5.4.2-4}
$$

and $m_1 = \rho h$ and $m_2 = \rho h^3/12$, ρ being the density. ϕ_x and ϕ_y denote the rotation of the transverse normal about the y and x-axis, and D_{11}, D_{12}, D_{22}, D_{66}, A_{44}, and A_{55} are the orthotropic plate stiffnesses

$$
D_{11}=\frac{E_1 h^3}{12(1-\nu_{12}\nu_{21})},\qquad D_{22}=\frac{E_2}{E_1}D_{11}
\tag{5.4.2-5}
$$

$$D_{12} = \nu_{12} D_{22} = \nu_{21} D_{11}, \quad D_{66} = \frac{1}{12} G_{12} h^3 \qquad (5.4.2\text{-}6)$$

$$A_{44} = G_{23} hk, \quad A_{55} = G_{13} hk \qquad (5.4.2\text{-}7)$$

where k is the shear correction coefficient taken to be 5/6.

Let the two-dimensional region Ω be subdivided into a number of finite elements $\Omega_e(e = 1,2,\ldots N)$. Over each element the generalized displacements (w, ϕ_x, ϕ_y) are interpolated according to the form

$$w_e(x,y,t) = \sum_{i=1}^{n} w_i^e(t) \psi_i^e(x,y) \qquad (5.4.2\text{-}8)$$

$$\phi_{x_e}(x,y,t) = \sum_{i=1}^{n} \phi_{x_i}^e(t) \psi_i^e(x,y) \qquad (5.4.2\text{-}9)$$

$$\phi_{y_e}(x,y,t) = \sum_{i=1}^{n} \phi_{y_i}^e(t) \psi_i^e(x,y) \qquad (5.4.2\text{-}10)$$

where ψ_i is the interpolation function corresponding to the *ith* node in the element. The interpolation function can be linear, quadratic, or cubic.

Substituting Equations 5.4.2-8 through 5.4.2-10 into the Galerkin integrals associated with the operator Equation 5.4.2-1, which must hold in each element Ω_e, gives

$$\int_{\Omega_e} ([L]\{\delta\} - \{f\})\{\psi_i\}\, dxdy = 0 \qquad (5.4.2\text{-}11)$$

where $\{\delta\} = \{w, \phi_x, \phi_y\}^T$. Integrating by parts once, yields an equation of the form

$$\overline{M}\ddot{U} + \overline{K}U = \overline{F} \qquad (5.4.2\text{-}12)$$

As for the one-dimensional beam problem, to reduce the order of the system, static condensation will be used. This condensation relates the dependent or secondary degrees of freedom (rotational) to the primary degrees of freedom (transverse). The relation between the secondary and primary degrees of freedom is found by establishing the static relation between them. This method is simple to apply, but is only an approximation and may produce large errors if care is not taken. The final reduced dynamic system is given by

$$\mathbf{M}\ddot{\mathbf{u}} + \mathbf{K}\mathbf{u} = \mathbf{F} \tag{5.4.2-13}$$

where **u** is a vector containing all the transverse displacements of the model, **M** is the assembled mass matrix, **K** is the assembled stiffness matrix, and **F** represents the applied forces. Equation 5.4.2-13 as a first-order system is written as

$$\begin{Bmatrix} \dot{x}_1 \\ \dot{x}_2 \end{Bmatrix} = \begin{bmatrix} \mathbf{0} & \mathbf{I} \\ -\mathbf{M}^{-1}\mathbf{K} & \mathbf{0} \end{bmatrix} \begin{Bmatrix} x_1 \\ x_2 \end{Bmatrix} + \begin{Bmatrix} 0 \\ \mathbf{M}^{-1}\mathbf{F} \end{Bmatrix} \tag{5.4.2-14}$$

or

$$\dot{\mathbf{x}} = \mathbf{K}^{*}\,\mathbf{x} + \mathbf{f} \tag{5.4.2-15}$$

These differential equations are converted to discrete equations using the exponential matrix as described in Chapter 2. The final discrete model is

$$\mathbf{x}_{j+1} = \mathbf{T}\mathbf{x}_j + \mathbf{P}\mathbf{f}_j \tag{5.4.2-16}$$

where **x** represents a vector of length 2n containing the displacements and velocities of the nodes. **T** is a matrix which represents the dynamics of the model, **f** is a vector of length n_g representing the unknown applied forces, and **P** is a matrix (2n $\times$ n_g) relating the forces to the system. Values for **T** and **P** are given by

$$\mathbf{T} = e^{\mathbf{K}^{*}h}, \quad \mathbf{P} = \mathbf{K}^{*-1}(\mathbf{T} - \mathbf{I}) \tag{5.4.2-17}$$

Alternately, for a backward implicit method the discrete equation is given by Equation 5.4.2-16 where

$$\mathbf{T} = \begin{bmatrix} \left(\mathbf{I} - \overline{\mathbf{K}}^{-1}\mathbf{K}h^2\right) & \overline{\mathbf{K}}^{-1}\mathbf{M}h \\ -\overline{\mathbf{K}}^{-1}\mathbf{K}h & \overline{\mathbf{K}}^{-1}\mathbf{M} \end{bmatrix}, \quad \mathbf{P} = \begin{Bmatrix} \overline{\mathbf{K}}^{-1}h^2 \\ \overline{\mathbf{K}}^{-1}h \end{Bmatrix} \tag{5.4.2-18}$$

and $\overline{\mathbf{K}} = \mathbf{M} + \mathbf{K}h^2$. In the same manner, the Newmark generalized acceleration method with $\delta = 1/2$ and $\alpha = 1/4$ yields

$$\mathbf{T} = \begin{bmatrix} \left(\mathbf{I} - \overline{\mathbf{K}}^{-1}\mathbf{K}\dfrac{h^2}{2}\right) & \left\{\mathbf{I} + \overline{\mathbf{K}}^{-1}\left(\mathbf{M} - \mathbf{K}\dfrac{h^2}{4}\right)\right\}\dfrac{h}{2} \\ -\overline{\mathbf{K}}^{-1}\mathbf{K}h & \overline{\mathbf{K}}^{-1}\left(\mathbf{M} - \mathbf{K}\dfrac{h^2}{4}\right) \end{bmatrix} \tag{5.4.2-19}$$

$$\mathbf{P} = \begin{Bmatrix} \overline{\mathbf{K}}^{-1}\dfrac{h^2}{2} \\ \overline{\mathbf{K}}^{-1}h \end{Bmatrix}, \quad \overline{\mathbf{K}} = \mathbf{M} + \mathbf{K}\frac{h^2}{4} \tag{5.4.2-20}$$

Again, suppose that a series of measurements have been taken and are replaced by the vector $\mathbf{d}_j$, where the length of $\mathbf{d}_j$ is m. The measurements could represent displacements, velocities, accelerations. These measurements are related to the state variables $\mathbf{x}_j$ by

$$\mathbf{d}_j \sim \mathbf{Q}\mathbf{x}_j \tag{5.4.2-21}$$

where $\mathbf{Q}$ is an $(m \times n)$ matrix which associates the measurements to the state variables.

The problem as stated in Section 5.3 is to find the unknown forces $\mathbf{f}_j$ that when they are used will force the model to best match the measurements.

5.4.3 Numerical Example

In order to demonstrate the proposed procedure the simply supported square orthotropic plate, with first-order shear deformation, shown in Figure 5.17 is employed. In our example, due to biaxial symmetry, only one quarter of the square simply supported plate was modeled with four quadratic elements. Thus, n = 16 in this example, but in practical problems n can easily reach 1000. The plate contains four nine-noded elements and has seventy-five degrees of freedom (DOF) (three DOF per node).

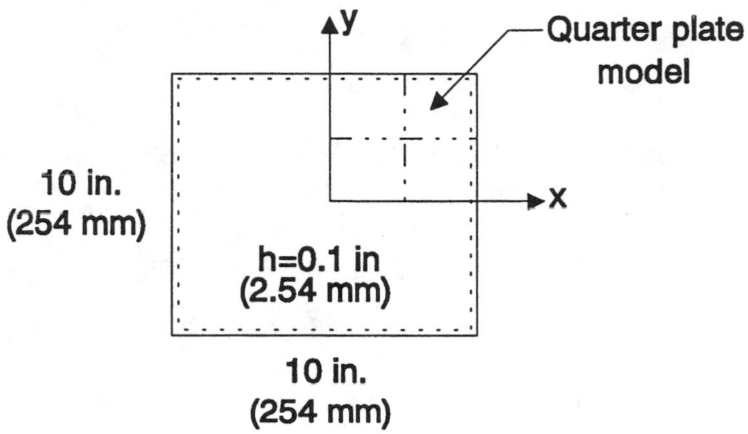

FIGURE 5.17
Simply supported square plate.

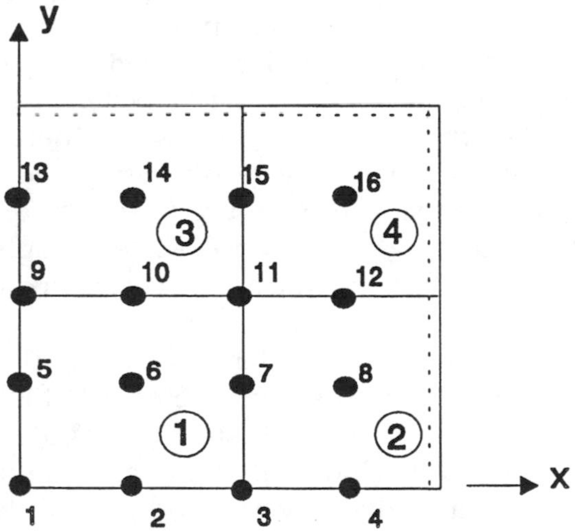

FIGURE 5.18
Finite element plate model.

However, the rotational degrees of freedom have been condensed out to reduce the order of the system. Thus, the final finite element model has sixteen transverse displacements as shown in Figure 5.18.

The orthotropic plate was assumed to be isotropic and the following properties were used:

$$E = E_1 = E_2 = 10.4 \times 10^6 \text{ psi. (72 MPa)}$$

$$G = G_{12} = G_{13} = G_{23} = 3.94 \times 10^6 \text{ psi. (27 MPa)}$$

$$\rho = 0.1 \text{ lb/in}^3 \text{ (2.8 Mg/m}^3\text{)}, \nu = 0.32$$

The load was applied at the center of the plate, and was a double rectangular pulse given by the following:

$$F = 50.0 \text{ lb. (222.4 N)}; \quad 0.005 < t < 0.0020; \quad 0.0035 < t < 0.005$$

$$F = 0.0; \quad 0 < t < 0.005; \quad t > 0.005$$

A direct solution of the model was used to solve for the displacement and velocity time histories. A timestep of 5.0E–5 seconds was used in the analysis. Using velocity as the measurement, a noise level using a standard deviation of 5 was added to the simulated measurement (see Figure 5.19). This represents approximately one percent of the peak value.

The zeroth-order regularization problem was solved using the exponential technique. Only the velocity measurement at node 11 at the edge of the plate was used in determining the unknown forcing function. The results using the exponential matrix technique are shown in Figure 5.20, where the b value was determined using GCV to be 5.8E–2. Although the results are noisy, they still reflect the true nature of the unknown impact force.

Replacing the regularization term with a derivative as was done for the cantilever beam gives the case of first-order regularization. The results using first-order regularization are shown in Figure 5.21. It is noted that the resulting impact force is much smoother and gives a very good representation of the true impact force. The optimal b value for this case was found to be 1.35.

In order to investigate the effect of different integration schemes, the experiment was repeated using Newmark and the backward implicit method.

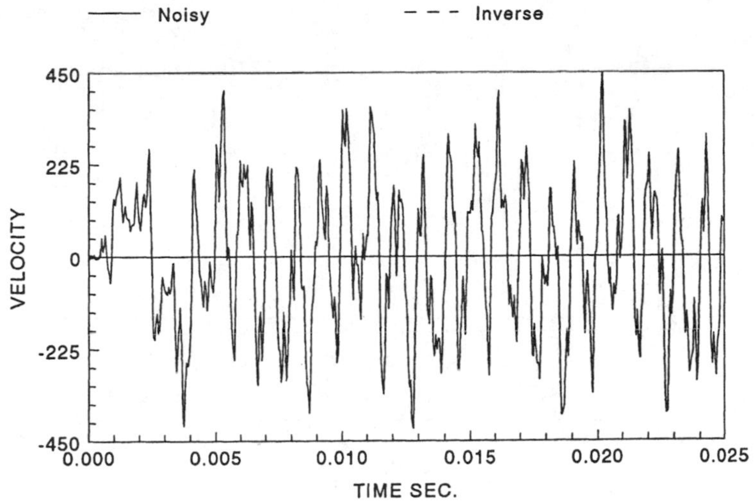

FIGURE 5.19
Velocity versus time.

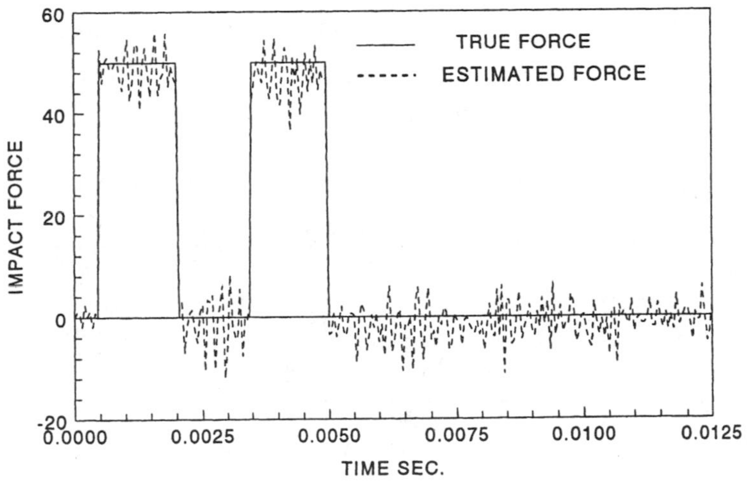

FIGURE 5.20
Impact force.

Newmark Method

The inverse problem for the plate under pulse loading was also investigated using the Newmark method described above. The results are shown in Figure 5.22 using first-order regularization. It is noted that this method tends to smooth out some

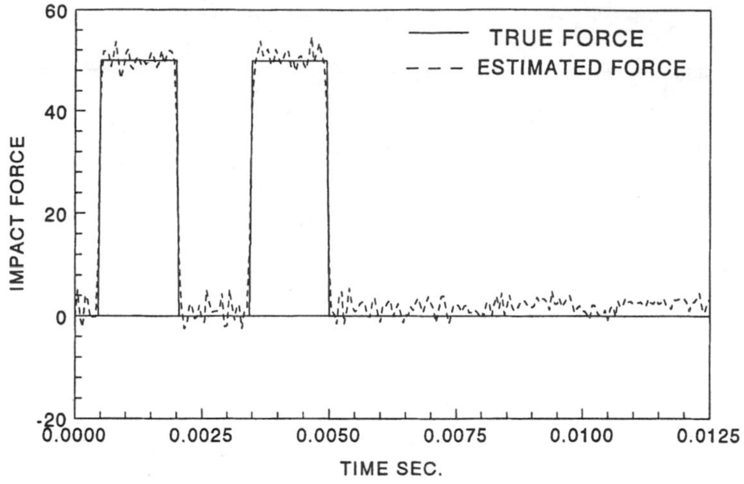

FIGURE 5.21
Impact force — first-order regularization.

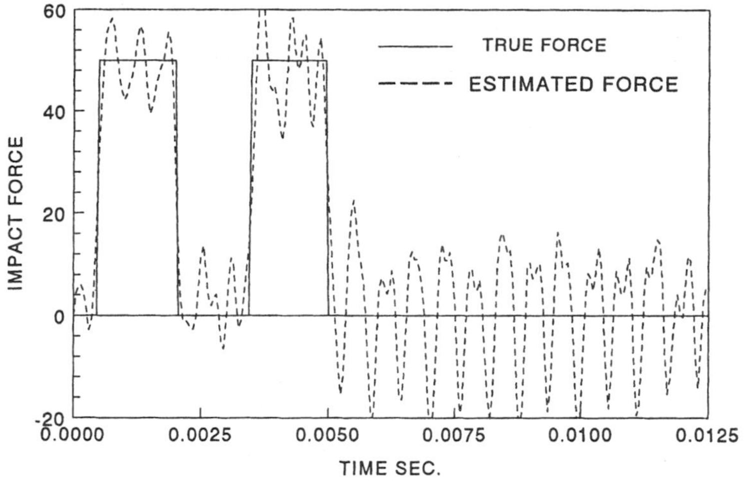

FIGURE 5.22
Impact force — Newmark method.

of the high frequency content of the forcing function. This is due to numerical damping. The regularization parameter b determined for this case by GCV was found to be 18.5.

Backward Implicit Method

In addition to the exponential method and the Newmark method, the backward implicit method was also used to evaluate

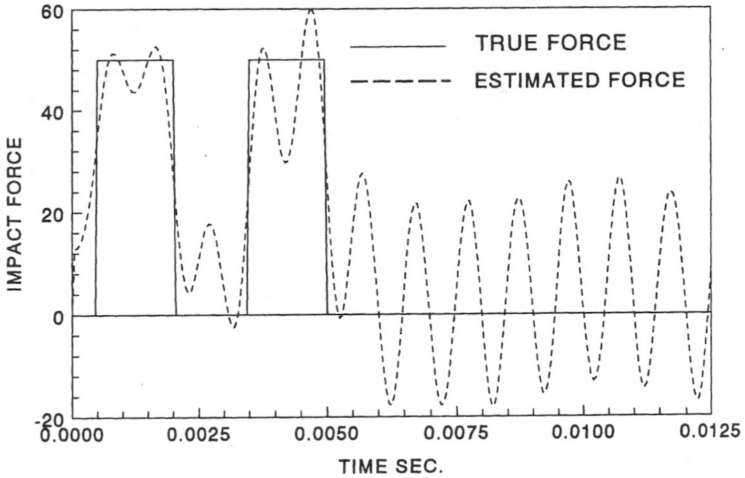

FIGURE 5.23
Impact force — backward implicit method.

the inverse pulse problem. The results for this case are shown in Figure 5.23. The backward implicit method tends to smooth out the high frequency content even more than that obtained from the Newmark method. The results were obtained using a first- order regularization. The value of b for this case was found to be 250.

The results of the above example give confidence that the exponential matrix along with generalized cross validation is a good method for solving the inverse structural dynamics problem. The method will work good for problems that have less than 1000 degrees of freedom. The Newmark and backward implicit methods can also be used, but are not as accurate as the exponential method. In addition, these methods have numerical damping, which also leads to inaccuracy.

Chapter 6

Smoothing and Differentiating Noisy Data

6.1 INTRODUCTION

The smoothing and differentiation of noisy data is a topic that fits naturally into the framework of the inverse methods developed in earlier chapters. First, it is an ill-conditioned problem in the sense that the smallest amount of noise in the data will tend to have a profound effect on the estimates of derivatives; and second, there is a wide range of models that can be used to smooth the data. In previous chapters we have been using physical models as the basis for developing smooth estimates of the variables. In this chapter we will show how to smooth data with the simplest of models that are not based on physical laws but on experience and judgment.

The smoothing of data is an intuitive process based on the assumption that most physical parameters are generally continuous and have continuous derivatives. The simplest case of fitting data with a straight line approximation is a perfect example of this assumption. And yet even this process can be somewhat subjective in that one might choose a second-order polynomial to fit the same data. The methodology presented in this chapter will not try to answer the difficult question of which model to use, although, if possible, it is good practice to rely on physical laws. But once a model is chosen, we will demonstrate how to match the data in an optimal manner.

The concepts used in smoothing range from the simplest least squares approach to digital filtering, Fourier series, and spline fitting. The approach in this chapter is to illustrate the general framework of dynamic programming and generalized cross validation as it applies to smoothing and differentiation of data. The use of generalized cross validation will play a vital role in our application in that it injects some objectivity into the smoothing process.

6.2 POLYNOMIAL APPROXIMATION

In Section 1.4 the simplest first-order system was used to smooth and differentiate some noisy data. In this section we will use a third-order model that will allow us to estimate the first and second derivatives of data. The continuous model is

$$\begin{aligned} \dot{x} &= v \\ \dot{v} &= a \\ \dot{a} &= g \end{aligned} \tag{6.2-1}$$

The model is to be driven by the forcing function g(t), and it also requires a set of initial conditions. This continuous model can be converted to a discrete one with the exponential matrix method of Section 2.3. In matrix form the equations are

$$\begin{bmatrix} \dot{x} \\ \dot{v} \\ \dot{a} \end{bmatrix} = \begin{bmatrix} 0 & 1 & 0 \\ 0 & 0 & 1 \\ 0 & 0 & 0 \end{bmatrix} \begin{bmatrix} x \\ v \\ a \end{bmatrix} + \begin{bmatrix} 0 \\ 0 \\ 1 \end{bmatrix} [g] \tag{6.2-2}$$

By assuming a constant g_j over a time interval t_j to t_{j+1} the discrete model becomes

$$\begin{bmatrix} x_{j+1} \\ v_{j+1} \\ a_{j+1} \end{bmatrix} = \begin{bmatrix} 1 & h & h^2/2 \\ 0 & 1 & h \\ 0 & 0 & 1 \end{bmatrix} \begin{bmatrix} x_j \\ v_j \\ a_j \end{bmatrix} + \begin{bmatrix} h^3/6 \\ h^2/2 \\ h \end{bmatrix} [g_j] \tag{6.2-3}$$

where h represents a uniform timestep, $t_{j+1} - t_j$.

These equations are equivalent to representing x(t) in an interval with a cubic polynomial of the form

$$x(t) = x_j + v_j(t - t_j) + a_j(t - t_j)^2/2 + g_j(t - t_j)/6 \tag{6.2-4}$$

The error term is a combination of matching the data d_j and the regularization of g_j

$$E = \sum_{j=1}^{N} (d_j - x_j)^2 + bg_j^2 \tag{6.2-5}$$

The smoothing problem is to find the forcing terms g_j and the initial conditions that minimize the error function E and to also determine the optimal value of the smoothing parameter b. Equation 6.2-3 gives the definitions of the matrices **M** and **P** for the general Equation 1.5-1. The **Q** matrix in this case is

$$\mathbf{Q} = \begin{bmatrix} 1 & 0 & 0 \end{bmatrix} \tag{6.2-6}$$

As an example of how the third-order filter would be used to smooth and estimate derivatives, consider the function

$$\begin{aligned} f(t) &= \sin(\pi t)/\pi^2 \qquad 0 \le t \le 4 \\ f(t) &= (t-4)/\pi \qquad 4 \le t \le 5 \end{aligned} \tag{6.2-7}$$

A constant sampling increment of 0.025 was used, and some simulated normally distributed noise with a sigma of 0.015 was added to the function. Figure 6.1 shows the original and the noisy functions.

Applying the inverse equations and using generalized cross validation to estimate the optimal value of the smoothing parameter b produced the results shown in Figure 6.2. The optimal parameter was estimated to be 8.8E–6. The noise has been entirely eliminated, and the estimated x_i's agree very well with the original signal. The first and second derivatives are shown in Figures 6.3 and 6.4, respectively. Except for a large deviation at the beginning, the estimates agree very well for the rest of the time. For other examples and some experiments with the

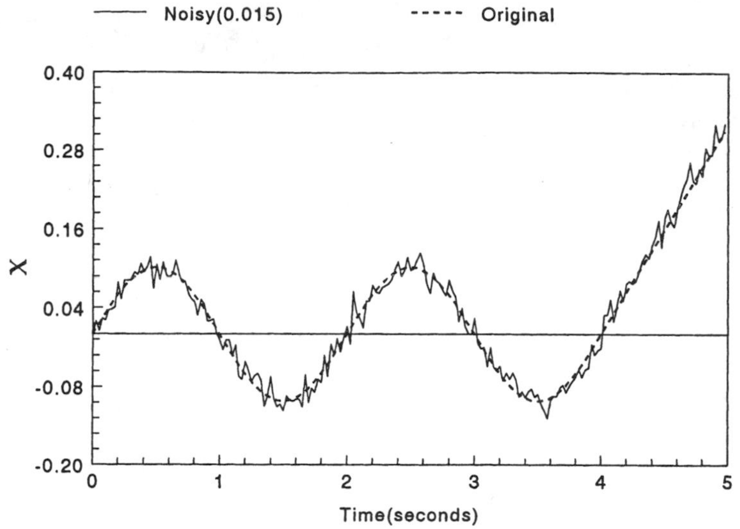

FIGURE 6.1
Noisy and original data.

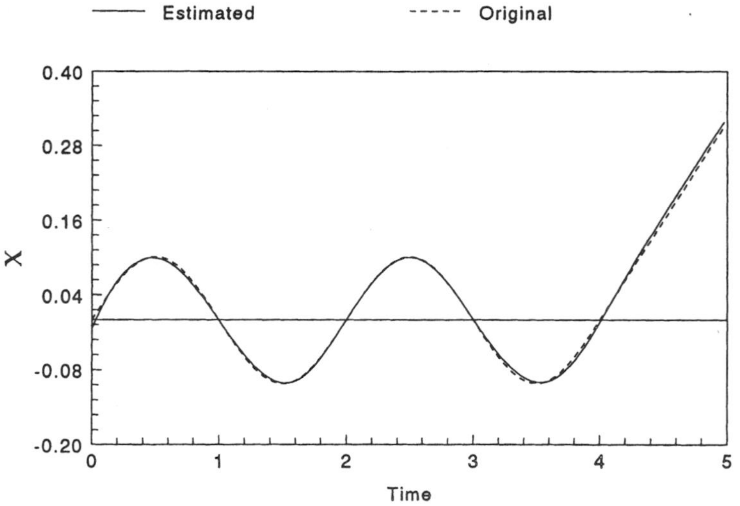

FIGURE 6.2
Comparison of estimated and original signal.

smoothing parameter, see Busby and Trujillo(1985) and Dohrmann, Trujillo, and Busby(1988).

The smoothing and integration of accelerometer data can also be performed with a similar formulation. See the works of

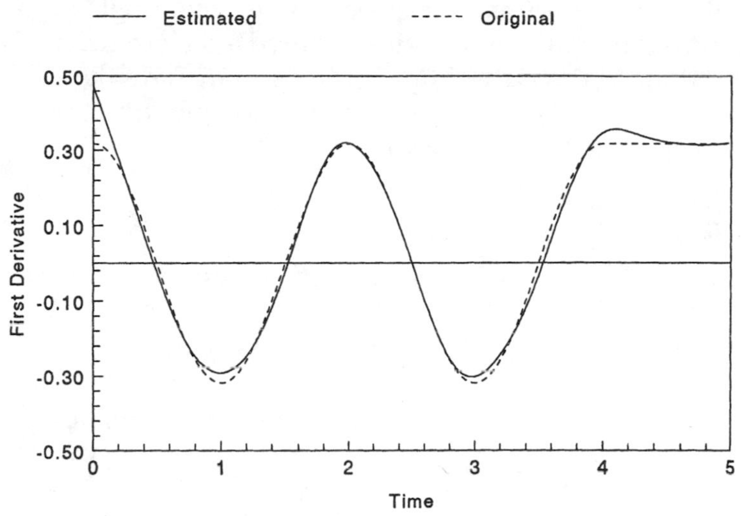

FIGURE 6.3
Comparison of estimated and original first derivative.

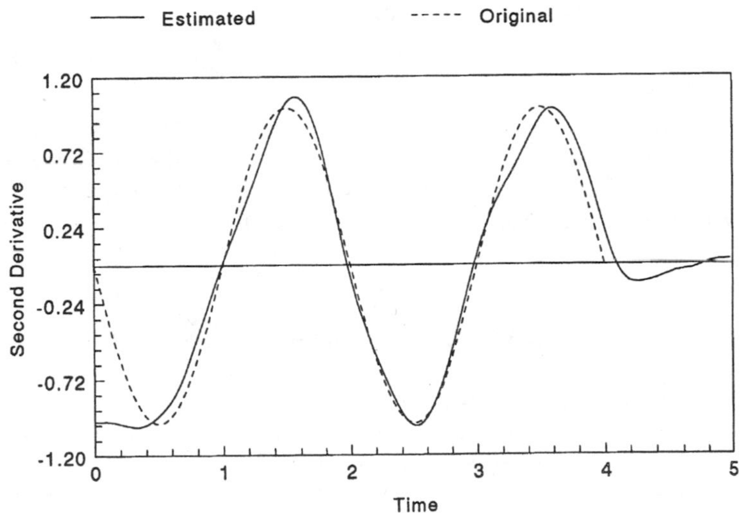

FIGURE 6.4
Comparison of estimated and original second derivative.

Trujillo and Carter (1982) and Murray (1985,1987). Another approach to analyzing accelerometer data would be to include the model of the accelerometer itself into the system, especially a seismic accelerometer. In cases like this, the instrument itself has distorted the signal and needs to be accounted for in the estimate. Some studies of skin-mounted accelerometers have shown that it is a fairly simple process to include the mass of the accelerometer in the model and subsequently correct the data (Trujillo and Busby, 1990).

Exercises

1. Derive Equation 6.2-3 using the matrix exponential formulas.
2. Derive Equation 6.2-3 using the polynomial formula in Equation 6.2-4.
3. The inverse method now involves two levels of optimization. For each b, an optimal forcing function is found. On another level, a function V(b) is minimized to estimate the optimal b. What criteria could be used to introduce even a third level of optimization which would involve the choice of models — first-order, second-order, etc.?
4. Is there any advantage to using only the steady-state Riccati matrix in the filter? See Murray (1985).
5. Reformulate the filter so that it can be programmed independent of the stepsize h.
6. Suppose one wanted to use a third-order polynomial over two data points at a time. This is equivalent to keeping g_i constant for two timesteps. Reformulate the dynamic programming formulas under these assumptions.

6.3 FILTERING A 60 HZ SIGNAL

One source of extraneous noise in measurements is the common 60 Hz interference. By adjoining a second-order differential equation to the third-order filter, it is possible to design a model that will eliminate both noise and any 60 Hz component from a signal. It is not usually possible to determine two unknown forcing functions from one measurement. In this case, however, it is the special restrictive nature of this model that makes this possible. The idea is to match the data d_j with a combination of x and y where these functions are determined from the following models: x is from a third-order filter

$$\begin{aligned}\dot{x} &= v \\ \dot{v} &= a \\ \dot{a} &= g\end{aligned} \tag{6.3-1}$$

and y from a second-order model of the form

$$\ddot{y} + (2\pi 60)^2 y = f \tag{6.3-2}$$

The models are to be driven by the unknown forcing functions g(t) and f(t). They also require a set of initial conditions. As usual, this continuous model can be converted to a discrete one with the exponential matrix method of Section 2.3. In matrix form the adjoined equations are

$$\begin{bmatrix}\dot{x}\\ \dot{v}\\ \dot{a}\\ \dot{y}\\ \dot{w}\end{bmatrix} = \begin{bmatrix}0 & 1 & 0 & 0 & 0\\ 0 & 0 & 1 & 0 & 0\\ 0 & 0 & 0 & 0 & 0\\ 0 & 0 & 0 & 0 & 1\\ 0 & 0 & 0 & -(2\pi 60)^2 & 0\end{bmatrix}\begin{bmatrix}x\\ v\\ a\\ y\\ w\end{bmatrix} + \begin{bmatrix}0 & 0\\ 0 & 0\\ 1 & 0\\ 0 & 0\\ 0 & 1\end{bmatrix}\begin{bmatrix}g\\ f\end{bmatrix} \tag{6.3-3}$$

The error term in this case is

$$E = \sum_{j=1}^{N} \left(d_j - x_j - y_j\right)^2 + b\left(g_j^2 + f_j^2\right) \tag{6.3-4}$$

As stated in the previous section, the smoothing problem is to find the forcing terms g_j and f_j and the initial conditions that minimize the error E. It is also necessary to estimate the optimal value of the smoothing parameter b. In the matrix notation of Section 1.5 the matrix **Q** is a (1 × 5) vector given by

$$\mathbf{Q} = \begin{bmatrix}1 & 0 & 0 & 1 & 0\end{bmatrix} \tag{6.3-5}$$

and the **B** matrix is a (2 × 2) identity matrix. The **M** matrix will be of size (5 × 5) and **P** is (5 × 2). A more general approach would be to introduce two independent smoothing parameters in the matrix **B** and to minimize the generalized cross-validation function over both of these. Another approach might be to use a single b but weight them differently in **B**. In this example only a single b with equal weights was used.

Since y is restricted by the differential equation to generate a 60 Hz signal it will tend to follow that part of the signal, leaving x, which is a smooth function, to follow the remainder. As an example of how this model performs, the function used in Section 6.2 was further corrupted with a signal of the form 0.015sin ($2\pi 60t$). The amplitude is equal to the value of the noise added to the original function given by Equation 6.2-6. The resulting function was sampled using a constant increment of 0.0025 for a total of 2000 data points. These data points are shown in Figure 6.5. The small sampling increment is necessary to assure a good sampling of the 60 Hz signal.

The generalized cross-validation estimation of the optimal smoothing parameter was b = 3.37E–6. The estimated and original signal are shown in Figure 6.6 and the first and second derivatives in Figure 6.7 and 6.8, respectively. The estimates are excellent and are slightly better than those obtained in Section 6.2. This is probably due to the effect of including more samples in the data.

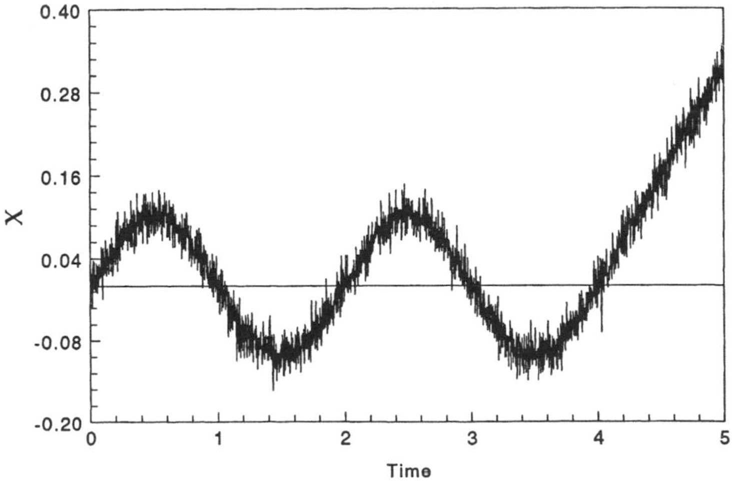

FIGURE 6.5
Original signal with noise and 60 Hz signal.

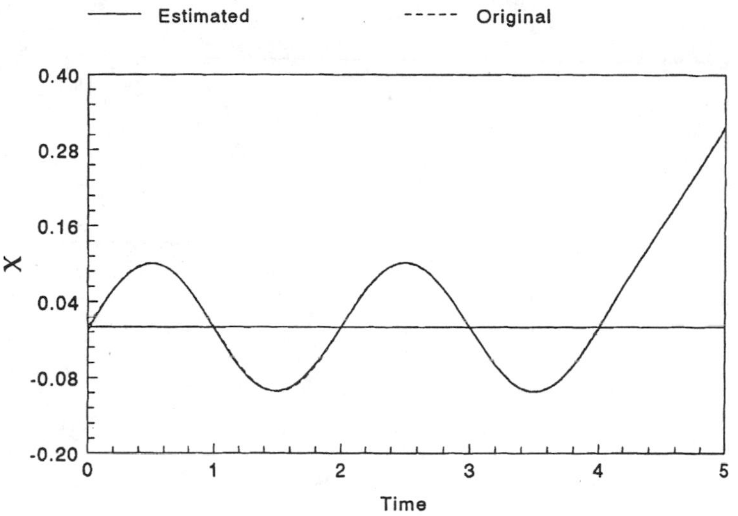

FIGURE 6.6
Comparison of estimated and original signal (60 Hz Example).

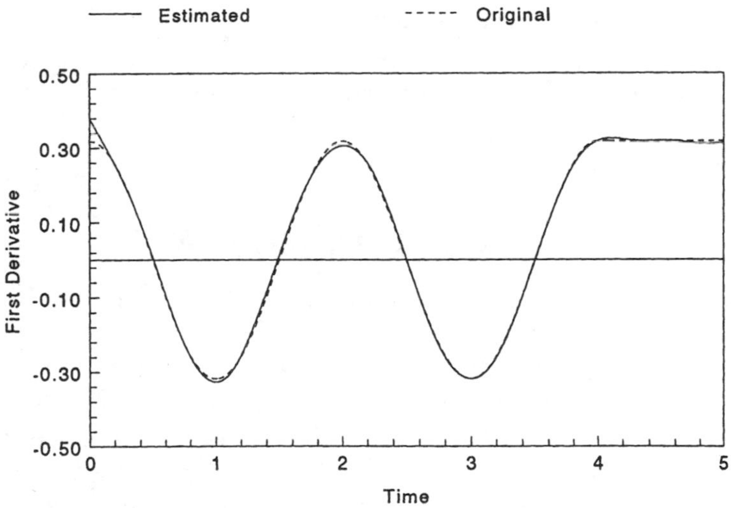

FIGURE 6.7
Comparison of estimated and original first derivative (60 Hz).

Exercises

1. Repeat the above example using sample sizes of 1500, 1000, and 500. In particular, monitor the ability of the methods to estimate the optimum smoothing parameter.

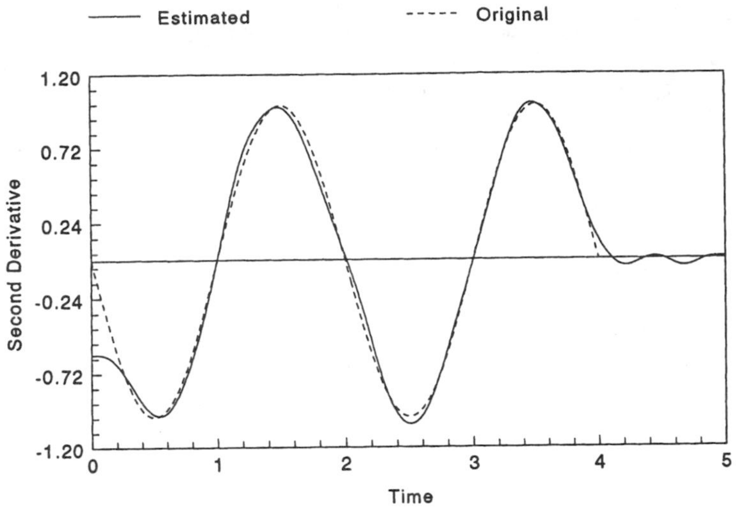

FIGURE 6.8
Comparison of estimated and original second derivative (60 Hz).

2. Repeat the above example with a case where the amplitude of the 60 Hz signal is not constant but varies slightly. This is an example of an adjustable filter.
3. What if the data were corrupted with a 59 Hz signal and the model was kept at 60 Hz? This investigates the sensitivity of the filter.

6.4 FREQUENCY ANALYSIS

For linear systems, a very useful way to describe the behavior of filters is with a frequency analysis. This analysis is a mathematical extension of what an experimenter might do to analyze a black box — which is to simply apply a sine wave of fixed amplitude and frequency to the input and observe the output. For a linear system, the output will also be a sine wave which may have a different amplitude and be out of phase with the input signal. This experiment is repeated with another frequency until a complete mapping of the transfer function is obtained. This type of frequency analysis is limited to a steady-state situation.

This experiment can also be performed mathematically. For ease of analysis the input sine wave is expressed in complex form as $e^{i\omega t}$, where ω is the frequency in radians/time. The output is also expressed in complex form as $A(\omega)e^{i\omega t}$ where $A(\omega)$ is the transfer function. Unlike most filters, the filters of interest have two sweeps, forward and backward, so care must be taken to insure that the steady-state response is obtained.

Consider a first-order filter

$$x_{k+1} = x_k + g_k \tag{6.4-1}$$

The smoothing and regularization method is to minimize the functional expressed in Equation 6.2-5. For this system, the dynamic programming solution can be expressed in terms of scalar equations. The backward sweep equations are

$$r_k = 1 + br_{k+1}/(b + r_{k+1}) \tag{6.4-2}$$

$$s_k = -2d_k + bs_{k+1}/(b + r_{k+1}) \tag{6.4-3}$$

and the forward sweep equations are

$$x_{k+1} = x_k - (s_{k+1} + 2r_{k+1}x_k)/2(b + r_{k+1}) \tag{6.4-4}$$

The steady-state response is obtained by letting the r_k's reach a steady-state value. This value can be found directly from Equation 6.4-2 and is given by

$$r = 0.5\left(1 + \sqrt{1+4b}\right) \tag{6.4-5}$$

It is clear that the steady-state value, and hence the transfer function, is dependent on the smoothing parameter b. The discrete data with a unit amplitude is now expressed as $d_k = e^{i\omega hk}$ where the sampling increment is h. The solution for s_k can be expressed as $s_k = se^{i\omega hk}$ and Equation 6.4-3 becomes

$$s = 2/\left[be^{i\omega h}/(b + r)^{-1}\right] \tag{6.4-6}$$

The steady-state response for x can be obtained by assuming that

$$x_k = Ae^{i\omega hk} \tag{6.4-7}$$

and using the steady-state values of s and r in Equation 6.4-4 gives, after some simplification,

$$A(\omega h, b) = 1/\left[1 + 2b\left(1 - \cos(\omega h)\right)\right] \tag{6.4-8}$$

The fact that $A(\omega h,b)$ is real means that there is no phase shift in the response x_k for all frequencies. A plot of $A(\omega h,b)$ is shown in Figure 6.9 for several values of the smoothing parameter b. The Nyquist sampling frequency is $\omega h = (2\pi f = \omega)$. It is clear from the figure that the cutoff frequency of the filter is shifted to lower values as b is increased. The relationship between the smoothing parameter b and the cutoff frequency means that determining the optimum value of b is equivalent to determining the *optimum cutoff frequency* of the filter for that particular signal.

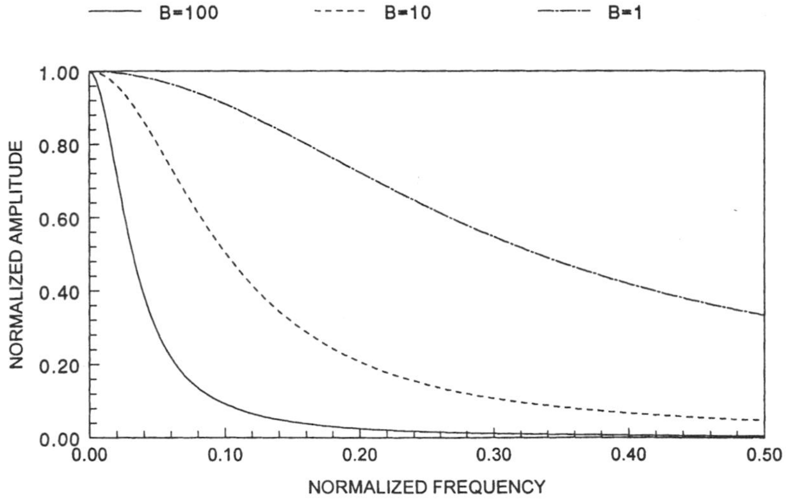

FIGURE 6.9
Transfer function versus frequency for first-order filter.

The frequency analysis for the third-order filter follows the same logical steps as for the first-order filter. The difference is that the algebra now involves 3 × 3 matrices and 3 × 1 vectors. The first step is to use Equation 1.5-22 to find the steady-state matrix **R**. Using this steady-state **R** gives a steady-state matrix $\underline{\mathbf{M}}^T$

$$\underline{\mathbf{M}}^{\mathrm{T}} = \mathbf{M}^{\mathrm{T}}\left(\mathbf{I} - \mathbf{H}^{\mathrm{T}}\mathbf{D}\mathbf{P}^{\mathrm{T}}\right) \tag{6.4-9}$$

The input to the system is $d_k = e^{i\omega hk}$. The steady-state value of the vector **s** is obtained from Equation 1.5-23

$$\mathbf{s}_{k-1} = -2\mathbf{Q}^{\mathrm{T}}\mathbf{d}_{k-1} + \underline{\mathbf{M}}^{\mathrm{T}}\mathbf{s}_k \tag{6.4-10}$$

The steady state is $\mathbf{s}_k = e^{i\omega hk}\mathbf{s}$ which gives

$$\mathbf{s} = 2\left(e^{i\omega h}\underline{\mathbf{M}}^{\mathrm{T}} - \mathbf{I}\right)^{-1}\mathbf{Q}^{\mathrm{T}} \tag{6.4-11}$$

The amplitudes can now be determined from the steady-state forward equation

$$\mathbf{x}_{j+1} = \underline{\mathbf{M}}^{\mathrm{T}}\mathbf{x}_J - \mathbf{P}\mathbf{D}\mathbf{P}^{\mathrm{T}}\mathbf{s}e^{i\omega h(j+1)} \tag{6.4-12}$$

where the vector **x** represents the three state variables of Equation 6.2-3. As before, the steady state is assumed to be $\mathbf{x}_j = e^{i\omega hj}\mathbf{x}$ which gives

$$\left(e^{i\omega h}\mathbf{I} - \underline{\mathbf{M}}\right)\mathbf{x} = -\mathbf{P}\mathbf{D}\mathbf{P}^{\mathrm{T}}\mathbf{s}e^{i\omega h} \tag{6.4-13}$$

or finally

$$\mathbf{x} = -2\left(e^{i\omega h}\mathbf{I} - \underline{\mathbf{M}}\right)^{-1}\left(\mathbf{P}\mathbf{D}\mathbf{P}^{\mathrm{T}}\right)\left(e^{i\omega h}\underline{\mathbf{M}}^{\mathrm{T}} - \mathbf{I}\right)^{-1}\mathbf{Q}^{\mathrm{T}}e^{i\omega h} \tag{6.4-14}$$

These expressions are too complicated to expand analytically. In order to produce the transfer functions, they were evaluated numerically by first calculating the steady-state values of $\underline{\mathbf{M}}$ and **D** and substituting them into Equation 6.4-14 and solving for **x**. Figures 6.10 through 6.12 show the transfer functions for various values of b. The displacement transfer function was real, which indicates a zero-phase angle throughout the frequency range. The shift in the cutoff frequency with the parameter b is clearly evident. The displacement transfer functions are quite similar to those for Butterworth filters. The similarities of smoothing with

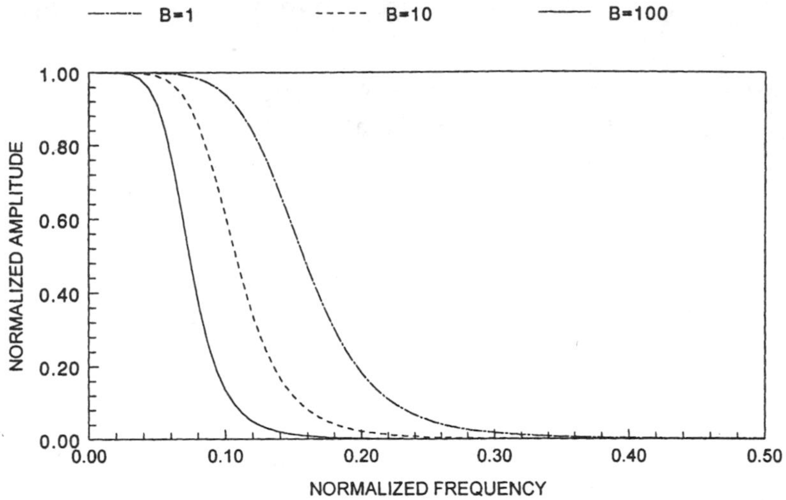

FIGURE 6.10
Transfer function versus frequency for third-order filter.

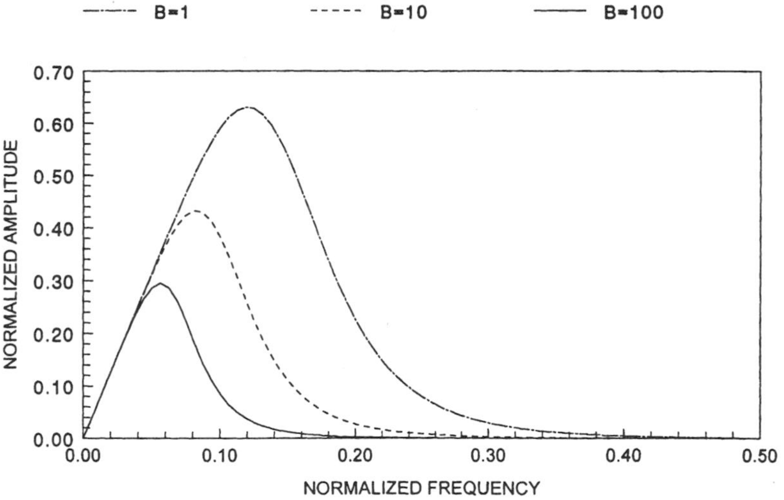

FIGURE 6.11
First derivative transfer function versus frequency, third order filter.

splines and Butterworth filters was first pointed out by Craven and Wahba (1979). A Butterworth filter is of thc form

$$A^2 = 1/\left[1+\left(f/f_0\right)^{12}\right] \tag{6.4-15}$$

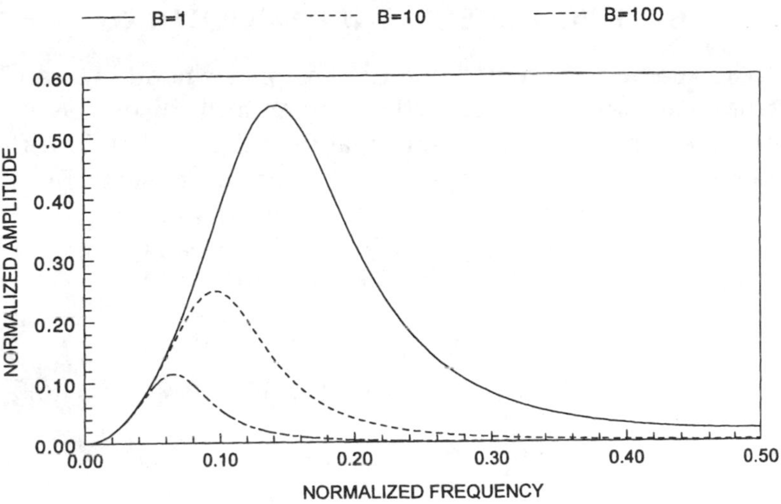

FIGURE 6.12
Second-derivative transfer function versus frequency, third order filter.

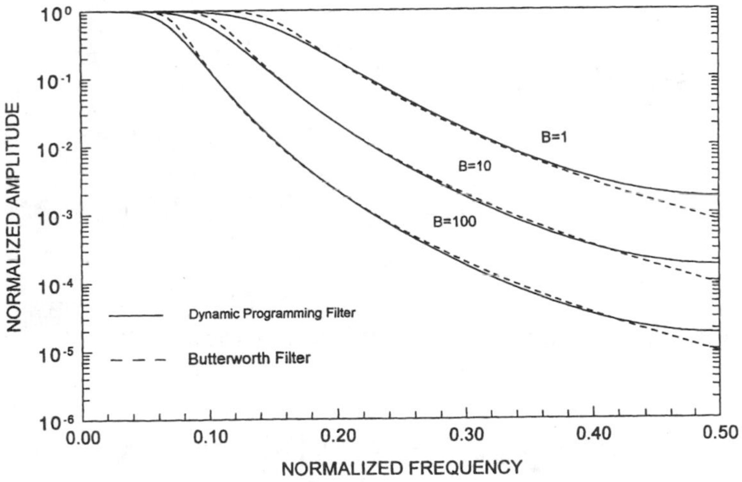

FIGURE 6.13
Transfer function — dynamic programming and Butterworth.

A plot of this transfer function is compared to the dynamic programming filter in Figure 6.13 for various values of the cutoff frequency f_0 and b. Once again it should be pointed out that the ability to determine the optimum smoothing parameter for any given signal is equivalent to determining the optimum cutoff frequency.

6.5 TWO-DIMENSIONAL SMOOTHING

The methods used for one-dimensional smoothing can be extended to two dimensions (Dohrmann and Busby, 1990). In order to extend the one-dimensional formulas, a slightly different expression for a cubic spline will be used. Consider the one-dimensional problem for the case of a uniform grid of $x_1, x_2, \ldots, x_n$, where $x_{i+1} = x_i + h$ and data d_i. Then f(x) can be represented as a combination of n + 2 basis functions

$$f(x) = \sum_{i=1}^{n+2} c_i s_i(x) \tag{6.5-1}$$

where

$$\begin{aligned} & s_1(x) = 1 \\ & s_2(x) = (x - x_1) \\ & s_3(x) = (x - x_2)^2/2 \\ & s_{i+3}(x) = \begin{cases} 0 & \text{for } x < x_i \\ (x - x_i)^3/6 & \text{for } x_i \le x < x_{i+1} \\ h^3/6 + h^2(x - x_{i+1})/2 + h(x - x_{i+1})^2/2 & \text{for } x \ge x_{i+1} \end{cases} \end{aligned} \tag{6.5-2}$$

The cubic spline is found by minimizing the functional

$$E(f) = \sum_{i=1}^{n} \left(f(x_i) - d_i\right)^2 + b \int_{x_1}^{x} \left(d^3f/dx^3\right)^2 dx \tag{6.5-3}$$

This formulation can be extended to two dimensions by assuming that the data d_{ij} are given on a uniform rectangular grid (x_i, y_i) where x_i is defined above and y_i satisfies $y_{i+1} = y_i + \underline{h}$. The basis functions for the y direction are defined similarly to those in the x direction.

$$
\begin{aligned}
&\underline{s}_1(y) = 1 \\
&\underline{s}_2(y) = (y - y_1) \\
&\underline{s}_3(y) = (y - y_2)^2/2 \\
&\underline{s}_{i+3}(y) = \begin{cases} 0 & \text{for } y < y_i \\ (y - y_i)^3/6 & \text{for } y_i \le y < y_{i+1} \\ \underline{h}^3/6 + \underline{h}^2(y - y_{i+1})/2 + \underline{h}(y - y_{i+1})^2/2 & \text{for } y \ge y_{i+1} \end{cases}
\end{aligned}
\tag{6.5-4}
$$

The bicubic spline function f(x,y) can now be expressed as a combination of products of basis functions in the x and y directions

$$
f(x,y) = \sum_{j=1}^{n+2} \sum_{i=1}^{n+2} c_{ij} s_i(x) \underline{s}_j(y) \tag{6.5-5}
$$

The bicubic spline and its derivatives are continuous over the entire rectangular grid because of the continuity of the basis functions and their derivatives.

The two-dimensional smoothing problem requires finding the bicubic spline that minimizes the functional

$$
\begin{aligned}
E(f) = &\sum_{j=1}^{n} \sum_{i=1}^{n} \left(f(x_i, y_j) - d_{ij}\right)^2 + b \sum_{j=1}^{n} \int_{x_1}^{x_n} \left(d^3f/dx^3\right)^2 dx \\
&+ b \sum_{i=1}^{n} \int_{y_1}^{y_n} \left(d^3f/dy^3\right)^2 dy + b^2 \int_{x_1}^{x_n} \int_{y_1}^{y_n} \left(d^6f/dx^3dy^3\right)^2 dxdy
\end{aligned}
\tag{6.5-6}
$$

Because of the special structure of Equation 6.5-6, the solution to the two-dimensional problem can be solved with a series of one-dimensional problems. This conclusion was inferred by examining the solution scheme outlined by Chiu and Schumaker, (1986). Basically, the procedure is to first smooth each row individually. These smoothed results are then further smoothed along each column. The columns could be smoothed first followed by the rows without affecting the solution. The generalized cross-validation function can be evaluated using the trace of the influence matrices in the x and y direction, Tr(**A**x) and Tr(**A**y). The final expression is

$$V(b) = \left(1/n^2\right)\sum_{j=1}^{n}\sum_{i=1}^{n}\left(f\left(x_i, y_j\right) - d_{ij}\right)^2 \Big/ \left[1 - \mathrm{Tr}(\mathbf{A}x)\,\mathrm{Tr}(\mathbf{A}y)/n^2\right]^2 \qquad (6.5\text{-}7)$$

If the amount of noise σ is known, Craven and Wahba (1979) then recommend that the smoothing parameter be chosen to minimize the functional

$$R(b) = \left(1/n^2\right)\sum_{j=1}^{n}\sum_{i=1}^{n}\left(f\left(x_i, y_j\right) - d_{ij}\right)^2 + 2\sigma^2\mathrm{Tr}(\mathbf{A}x)\,\mathrm{Tr}(\mathbf{A}y)/n^2 - \sigma^2 \qquad (6.5\text{-}8)$$

In order to demonstrate the performance of these functions, a simple example is presented. The function S(b), the error based on the original noise-free function, will also be compared.

The original function is

$$g(x, y) = \sin(\pi x/10)\sin(\pi y/10) \qquad (6.5\text{-}9)$$

which is sampled on the interval between 0 and 10 with a sampling increment of 0.2 for a total of 51 points in each direction. Some noise ($\sigma = 0.05$) was added to g(x,y) to simulate real data. This noisy data together with g(x,y) and the estimate f(x,y) are shown in Figure 6.14 for the slice y = 3. As the figure shows, the two-dimensional filter has produced an excellent estimate.

Since the exact function in this example has continuous derivatives, it is possible to compare these with those estimated

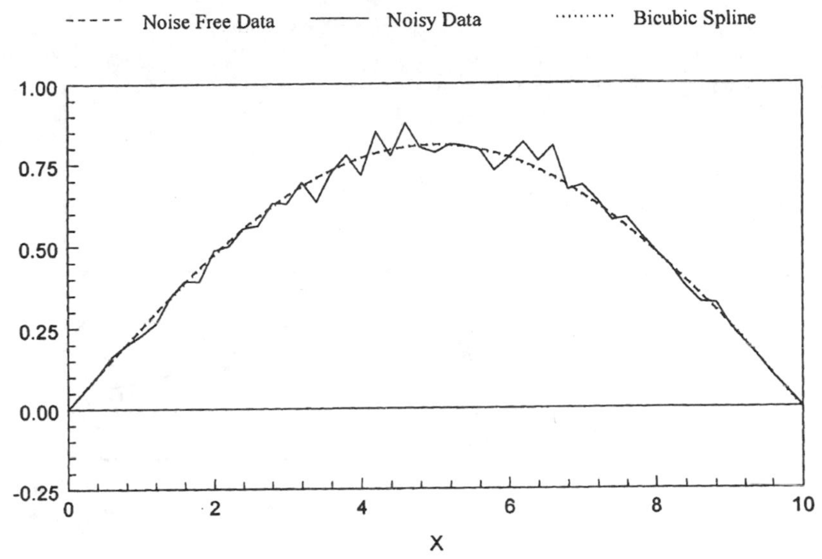

FIGURE 6.14
Comparison of noisy data — g and f along y = 3.

by the two-dimensional filter. Figure 6.15 compares the first partial derivative with respect to x along the y = 3 line. Similarly, Figure 6.16 compares the first partial derivative with respect to y. Since this is a two-dimensional function it also has cross derivatives. Figure 6.17 compares the second cross derivative along y = 3. In all three figures, the comparisons are excellent.

A second example is based on the function

$$g(x, y) = \sin(r)/r$$

which has been sampled on the interval $-3 < x,y < 3$ with a sampling increment of 0.12. The noise-free data are presented in Figure 6.18 and the noisy data ($\sigma = 0.02$) in Figure 6.19. Again the cross-validation function was used to select the smoothing parameter, which turned out to be 5.6389. The cross-validation function V(b) together with S(b) and R(b) (Equation 6.5-8)) are shown in Figure 6.20. The estimated function is shown in Figure 6.21. The agreement is excellent.

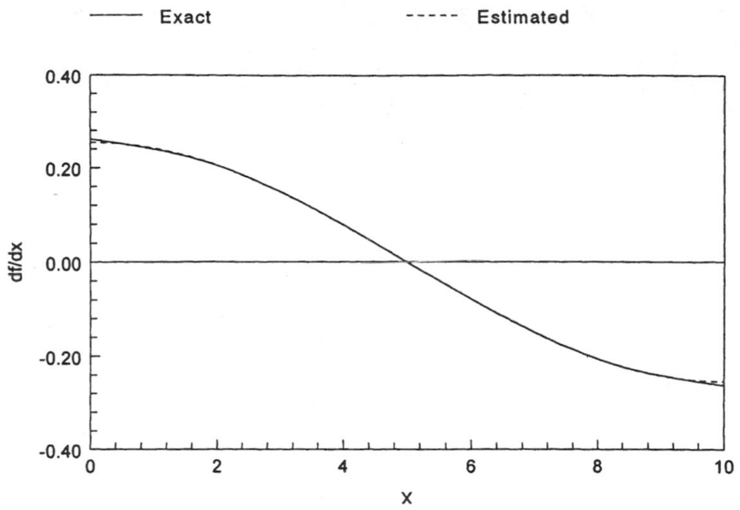

FIGURE 6.15
Comparison of exact and estimated first partial x derivative.

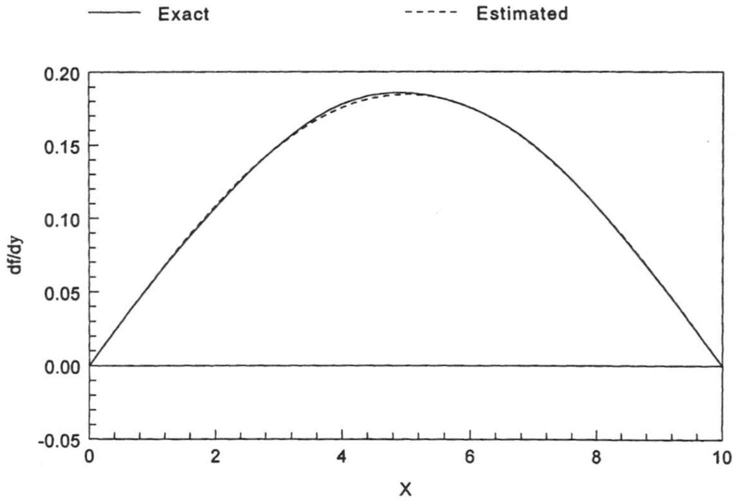

FIGURE 6.16
Comparison of exact and estimated first partial y derivative.

The decoupling of the computations into the x and y directions lends itself to a very efficient two-dimensional filter, espe-

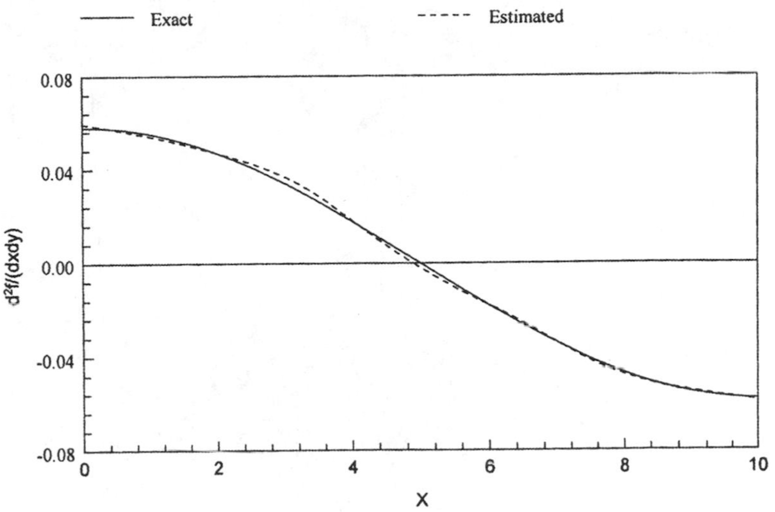

FIGURE 6.17
Comparison second cross partial derivative.

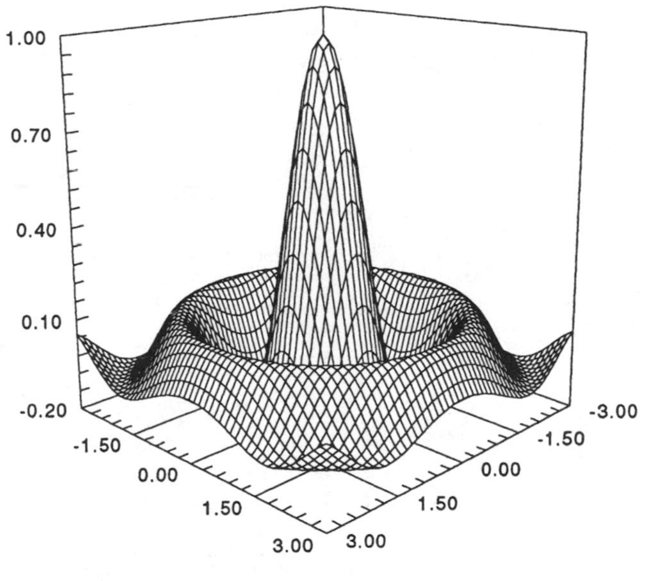

FIGURE 6.18
Noise-free data of Example 2.

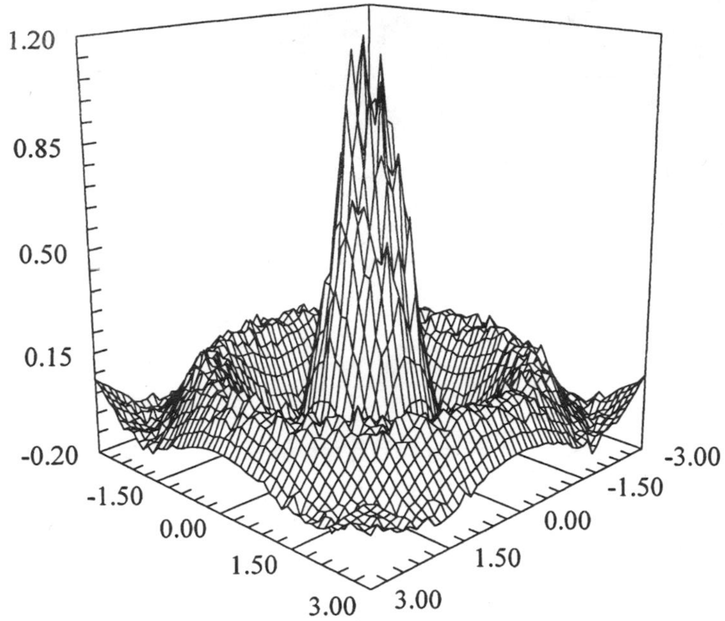

FIGURE 6.19
Noisy data of Example 2; sigma = 0.02.

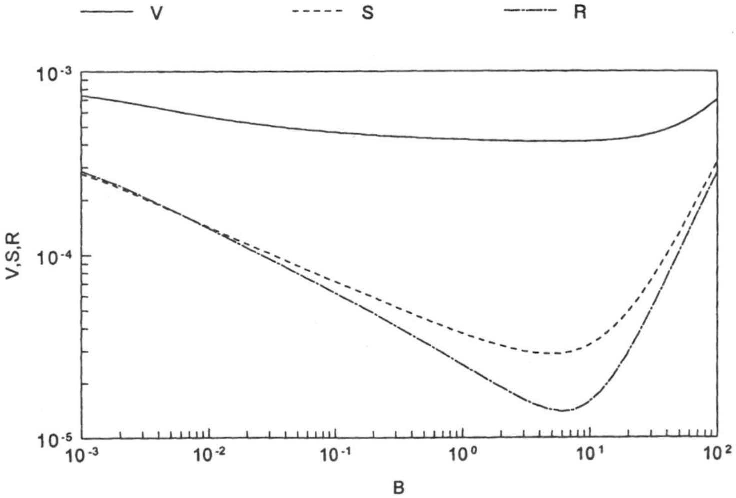

FIGURE 6.20
Plot of V, S, and R Example 2 versus the smoothing parameter.

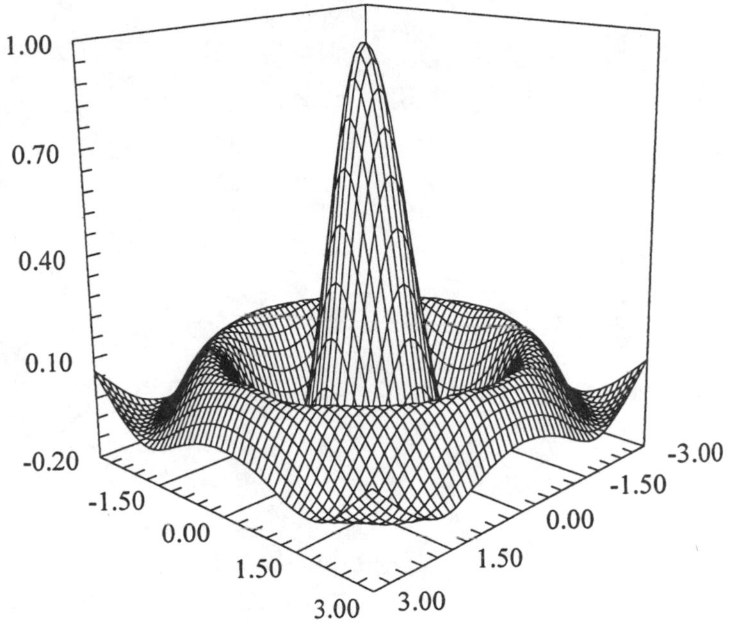

FIGURE 6.21
Estimated data using two-dimensional filter; b = 5.6389.

cially if a uniform spacing is used. In fact, many of the calculations performed during one backward sweep can be stored and reused for all the other rows and columns. Thus, in this case, the computations are proportional to the total number of points, giving a very efficient method.

Chapter 7

NONLINEAR SYSTEMS

7.1 INTRODUCTION

As one constructs models to better reflect true physical behavior, it becomes necessary to include nonlinear terms in the equations representing the models. The scope of nonlinear analysis is such that it cannot be encompassed in one general framework but must instead be discussed on a case-by-case basis. There is, however, a class of nonlinear models that can be analyzed with a common approach. These are models that are only slightly nonlinear. The common approach is to solve a sequence of linear problems such that each solution gets closer to the nonlinear one. Newton's method of solving a set of nonlinear equations is an example of such an approach. The success of such methods depends on two factors. One is that the linearized portion of the model dominate the nonlinear terms. The other is to start with a good first solution. It is beyond the scope of this book to attempt to quantify these factors. Usually one has sufficient knowledge of a system to help in the linearization and also to provide a good estimate of the solution.

Part of this chapter will present some results of inverse problems in the area of nonlinear heat conduction. The nonlinearities arise because the thermal properties depend on temperature. It is assumed that this nonlinearity is slight so that an appropriate linearized model can be used. There are several advantages to using a linearized model. First, all of the solution techniques presented in the previous chapters for linear systems can be used, including those that make possible very efficient computations. In particular, the Chandrasekhar equations allow us to analyze very large models. This makes the inverse method a practical tool for realistic models.

For some interesting analysis of nonlinear systems see Bellman (1973) and Distefano (1974). An excellent presentation of methods for nonlinear control can be found in Bryson and Ho (1975). The usual successful approach to nonlinear problems is to take advantage of all the knowledge garnered about a system and to experiment with different assumptions and methods.

7.2 LINEARIZATION METHODS

In linearizing a system of nonlinear equations for the inverse problem, it is required both to replace a continuous system with a discrete one and also to construct a framework that will allow the use of the linear inverse formulas. Consider the general nonlinear system of equations

$$\dot{\mathbf{x}} = \mathbf{f}(\mathbf{x}) + \mathbf{g}(t) \tag{7.2-1}$$

If only the direct problem was of interest, the above equation could be approximated with a simple explicit method

$$\mathbf{x}_{j+1} = \mathbf{x}_j + h\mathbf{f}(\mathbf{x}_j) + h\mathbf{g}_j \tag{7.2-2}$$

Comparing this with the general equation 1.5.3-1, it would be easy to identify the matrices **M** and **P**. All the nonlinear terms would be lumped into $\mathbf{v}_j$. However this approximation would not yield a useful model for the iterative solution of the inverse problem. A better approximation would be to further replace the nonlinear terms with a generalized Taylor's approximation. That is, let

$$\mathbf{f}(\mathbf{x}) = \mathbf{f}(\mathbf{x}^*) + [\partial\mathbf{f}/\partial\mathbf{x}^*](\mathbf{x} - \mathbf{x}^*) \tag{7.2-3}$$

where the term in brackets is the Jacobian matrix. The immediate question is what to choose for $\mathbf{x}^*$? For the nonlinear inverse problem, it will be assumed that $\mathbf{x}^*$ will be chosen from the results of a previously solved inverse problem. The model equation now becomes

$$\dot{\mathbf{x}} = \mathbf{f}(\mathbf{x}^*) + [\partial \mathbf{f}/\partial \mathbf{x}^*](\mathbf{x} - \mathbf{x}^*) + \mathbf{g} \qquad (7.2\text{-}4)$$

or

$$\dot{\mathbf{x}} = [\partial \mathbf{f}/\partial \mathbf{x}^*]\mathbf{x} + \mathbf{f}(\mathbf{x}^*) - [\partial \mathbf{f}/\partial \mathbf{x}^*]\mathbf{x}^* + \mathbf{g} \qquad (7.2\text{-}5)$$

This model can be converted to a discrete one using the simple implicit formula (see Table 2.1). First, let

$$\mathbf{A}_j = [\partial \mathbf{f}/\partial \mathbf{x}_j^*] \qquad (7.2\text{-}6)$$

$$\mathbf{v}_j^* = \mathbf{f}(\mathbf{x}_j^*) - [\partial \mathbf{f}/\partial \mathbf{x}_j^*]\mathbf{x}_j^* \qquad (7.2\text{-}7)$$

then Equation 7.2-5 becomes

$$\mathbf{x}_{j+1} = [\mathbf{I} - \mathbf{A}_j h]^{-1}\mathbf{x}_j + h[\mathbf{I} - \mathbf{A}_j h]^{-1}\mathbf{v}_j^* + h[\mathbf{I} - \mathbf{A}_j h]^{-1}\mathbf{g}_j \qquad (7.2\text{-}8)$$

This corresponds to the general form of Equation 1.5.3-1

$$\mathbf{x}_{j+1} = \mathbf{M}_j\mathbf{x}_j + \mathbf{v}_j + \mathbf{P}_j\mathbf{g}_j \qquad (7.2\text{-}9)$$

The main reason for choosing this approximation is that all of the $\mathbf{M}_j$'s and $\mathbf{v}_j$'s must be available when computing the backward sweep portion of the generalized inverse problem. The new $\mathbf{x}_j$'s are then computed during the forward sweep.

This simple approximation depends on an initial guess for $\mathbf{x}_j^*$'s. One possible way to generate these is to guess at the initial forcing terms $\mathbf{g}_j$'s. Another is to choose a representative constant value for all the $\mathbf{x}_j$'s and evaluate the nonlinear terms with these. Clearly, these approximations will not work for all nonlinear problems, but fortunately most models are only slightly nonlinear.

Another useful approximation is to keep the Jacobian matrix constant throughout the iterations. This tends to slow down the convergence, but each iteration is now much more computationally efficient. When used in the solution of nonlinear equations, this type of approximation is called a *modified Newton's method*. For example consider the model with the special form

$$\mathbf{f}(\mathbf{x}) = \mathbf{A}(\mathbf{x})\mathbf{x} \tag{7.2-10}$$

That is, the matrix **A** depends on **x**. In this case Equation 7.2-3 can be written as (ignoring higher derivatives)

$$\mathbf{f}(\mathbf{x}) = \mathbf{A}(\mathbf{x}^*)\mathbf{x}^* + \mathbf{A}(\mathbf{x}^*)(\mathbf{x} - \mathbf{x}^*) \tag{7.2-11}$$

Now, as a further approximation, let us set the second Jacobian matrix to a constant $\mathbf{A}_0$. Equation 7.2-8 now becomes

$$\begin{aligned}\mathbf{x}_{j+1} &= \left[\mathbf{I} - \mathbf{A}_0 h\right]^{-1}\mathbf{x}_j + h\left[\mathbf{I} - \mathbf{A}_0 h\right]^{-1}\left[\mathbf{A}\left(\mathbf{x}_j^*\right)\mathbf{x}_{j+1}^* - \mathbf{A}_0\mathbf{x}_{j+1}^*\right] \\ &+ h\left[\mathbf{I} - \mathbf{A}_0 h\right]^{-1}\mathbf{g}_j\end{aligned} \tag{7.2-12}$$

The advantage of this approximation is that the matrices **M** and **P** are constant, which means that the Chandrasekhar formulas can be used. Clearly in this case it will be easy to evaluate a constant matrix $\mathbf{A}_0$ with which to begin the process.

The inverse problem has one additional feature that will help in the convergence and that is the fact that the forcing terms will be calculated to always follow the data. This means, if the data are representative of the state variables, each solution will always be close to the data. In this case the largest variation during the iterations will be in the successive forcing terms $\mathbf{g}_j$.

The iteration process now involves solving successive inverse problems yielding the solutions $\mathbf{x}_j^*$ which are used in the next solution. Each inverse problem gives an estimate of the unknown forcing terms $\mathbf{g}_j$.

Exercise

1. Repeat the derivation for a Crank–Nicolson integration method.
2. Is Equation 7.2-12 consistent? That is, does it revert to the original differential equation as h approaches zero?

7.3 NONLINEAR INVERSE HEAT CONDUCTION

The Crank–Nicolson formulas used to integrate the direct equations for a nonlinear heat conduction model are

$$(\mathbf{C}_i + \mathbf{K}_i h/2)\mathbf{x}_{i+1} = (\mathbf{C}_i - \mathbf{K}_i h/2)\mathbf{x}_i + h\mathbf{q}_i + h\mathbf{P}\mathbf{f}_i \tag{7.3-1}$$

where $\mathbf{x}$ represents the temperatures, $\mathbf{K}_i$ is a symmetric conductance matrix containing the mathematical approximations to the spatial variables, $\mathbf{C}_i$ is the capacitance matrix, $\mathbf{q}_i$ is a vector representing the known heat fluxes, $\mathbf{f}_i$ is a vector representing the unknown heat fluxes, $\mathbf{P}$ is a participation matrix identifying the specified application of the unknown fluxes, and h is the integration timestep.

The thermal properties are assumed to vary with temperature, which means that the matrices $\mathbf{C}_i$ and $\mathbf{K}_i$ vary with temperature. In the direct problem this poses no difficulty since the previous temperatures $\mathbf{x}_i$ are known at each timestep and it is relatively easy to evaluate $\mathbf{C}_i$ and $\mathbf{K}_i$ using $\mathbf{x}_i$. However, for the inverse problem these matrices must be available for the backward sweep and must therefore be evaluated beforehand. Also, in order to use the Chandrasekhar equations it is useful to linearize with constant matrices. Using a *constant* $\mathbf{C}$ and $\mathbf{K}$, chosen appropriately, the above equation may be rewritten as

$$(\mathbf{C} + \mathbf{K}h/2)\mathbf{x}_{i+1} = (\mathbf{C} - \mathbf{K}h/2)\mathbf{x}_i + \mathbf{q}_i^* + h\mathbf{P}\mathbf{f}_i \tag{7.3-2}$$

where the nonlinear correction term $\mathbf{q}_i^*$ is

$$\mathbf{q}_i^* = h\mathbf{q}_i + h(\mathbf{K} - \mathbf{K}_i)/(\mathbf{x}_{i+1} + \mathbf{x}_i^*)/2 + (\mathbf{C} - \mathbf{C}_i)(\mathbf{x}_{i+1}^* - \mathbf{x}_i^*) \tag{7.3-3}$$

$\mathbf{C}_i$ and $\mathbf{K}_i$ are evaluated using the temperatures $\mathbf{x}_i^*$ from the previous inverse solution. The first solution is obtained by using the constant matrices $\mathbf{C}$ and $\mathbf{K}$.

Thus, the nonlinear inverse heat conduction problem is an iterative one. The steps are:

Step 1 Select a constant $\mathbf{C}$ and $\mathbf{K}$ and solve the inverse problem using only the known heat fluxes $\mathbf{q}_i$. This generates temperatures $\mathbf{x}^*$.

Step 2 Use the temperatures $\mathbf{x}_i^*$ to evaluate the nonlinear correction terms $\mathbf{q}_i^*$ using Equation 7.3-3.

Step 3 Solve the inverse problem with the correction terms $\mathbf{q}_i^*$.

Recall that the principle of superposition outlined in Section 1.5.3 must be used with $\mathbf{q}_i^*$ as a known forcing term.

Steps 2 and 3 are repeated until the unknown heat fluxes $\mathbf{f}_i$ converge. Since each inverse solution will tend to follow the data, it is also useful to monitor the temperatures for convergence. The above iterative process has been very successful in solving quenching problems (see Trujillo and Wallis, 1989). The next sections include some numerical examples of the above iterative process.

7.3.1 One-Dimensional Example

This one-dimensional problem represents a rod of length 4 divided into 40 equal segments. One end is insulated and an unknown heat flux is applied at the other. The temperature measurement is taken at x = 0.5. The capacitance is constant at 0.0283, but the thermal conductivity k varies linearly with temperature as shown below

Temperature	k
0	6.0E–4
1000	3.0E–4

The rod is initially at 1000 degrees and the measurements were taken every 0.5 seconds. The constant matrices **C** and **K** of Equation 7.3-2 were evaluated at the initial temperature.

For this example, the measurements were obtained by solving a direct problem using a heat flux that was equal to –0.40 between times 10 and 40 and zero elsewhere. In order to investigate only the effects of the nonlinearities, the measurements were not corrupted with noise. The results are shown in Figure 7.1 as the estimated heat flux histories for each iteration. A smoothing parameter equal to 100 was used. The convergence of the heat flux histories is evident in the figure. As expected for this type of nonlinear problem, the initial heat flux estimate using only the constant properties produced a fairly reasonable first estimate. The rapid convergence during the early time is due to

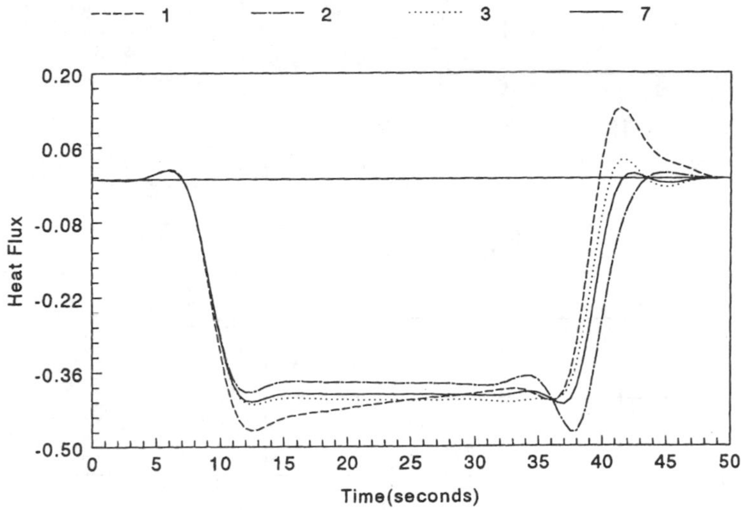

FIGURE 7.1
Heat flux estimations at various iterations.

the fact that the model is close to the initial temperature of 1000 degrees, which is the temperature used to evaluate the constant matrices. At later times the temperature of the model is much less than 1000, which produces a larger nonlinear correction term $\mathbf{q}_i^*$. In all of the iterations the measurements were followed very closely by the estimated temperatures.

It should be pointed out that besides the advantage of being able to use the Chandrasekhar equations with the constant matrices, there is an additional computational advantage of having to compute the backward Riccati matrices only once. Thus after the first iteration, only Equation 3.4-19 has to be solved during the backward sweep. This requires storing some matrices, which can easily be done.

7.3.2 Two-Dimensional Examples

The first two-dimensional example represents an axisymmetric body with three unknown heat fluxes and five measurements. The model is shown in Figure 7.2 together with the location of the measurements. In this example both the capacitance and the thermal conductivity vary with temperature as shown below.

The model is initially at 2200 degrees and the timestep is 0.5. The measurements were simulated by solving a direct problem

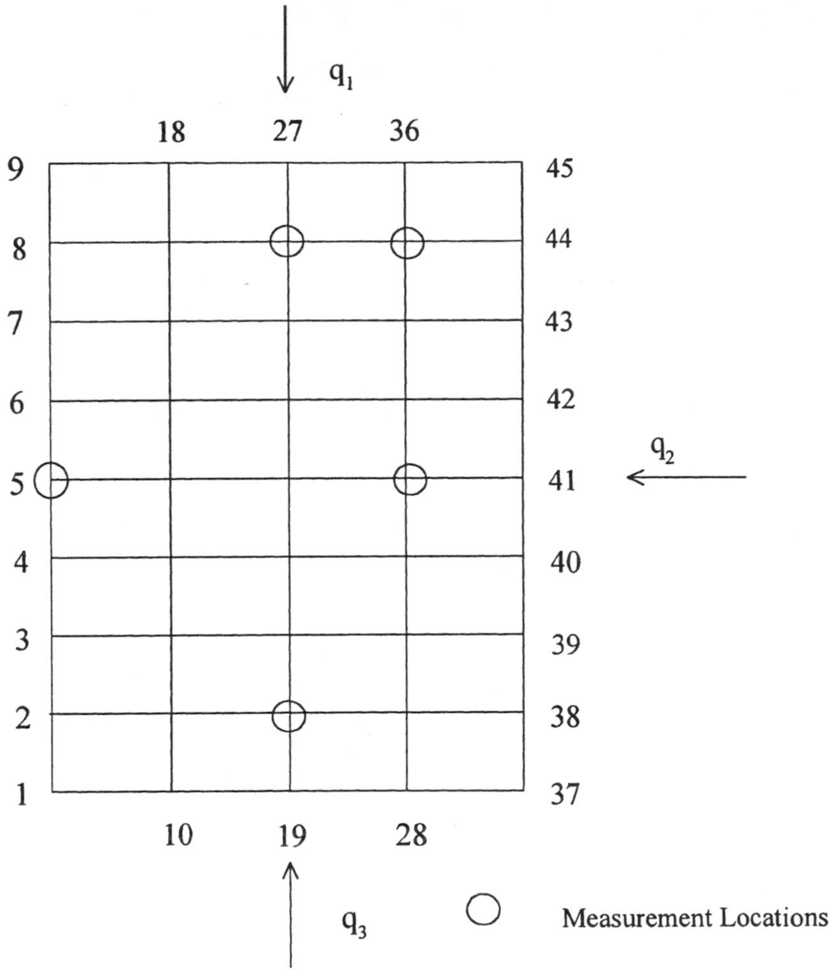

FIGURE 7.2
Finite element model, two-dimensional example.

Temperature	k	c
30	1.738E–4	3.015E–2
2200	3.289E–4	5.180E–2

using known heat fluxes applied to the top, bottom, and the outside of the model. The measurements were truncated to one degree to more closely represent real data. The results are shown in Figures 7.3 through 7.5 for the three heat fluxes, respectively. The convergence was very similar to that of the one-dimensional

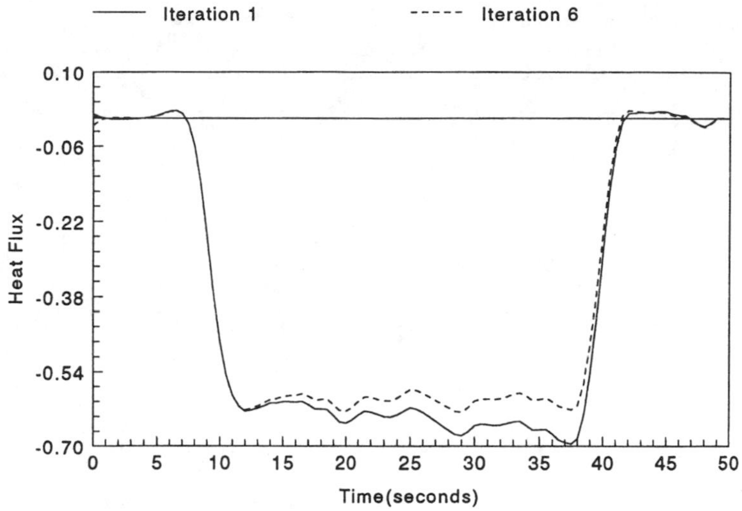

FIGURE 7.3
Heat flux 1, two-dimensional example.

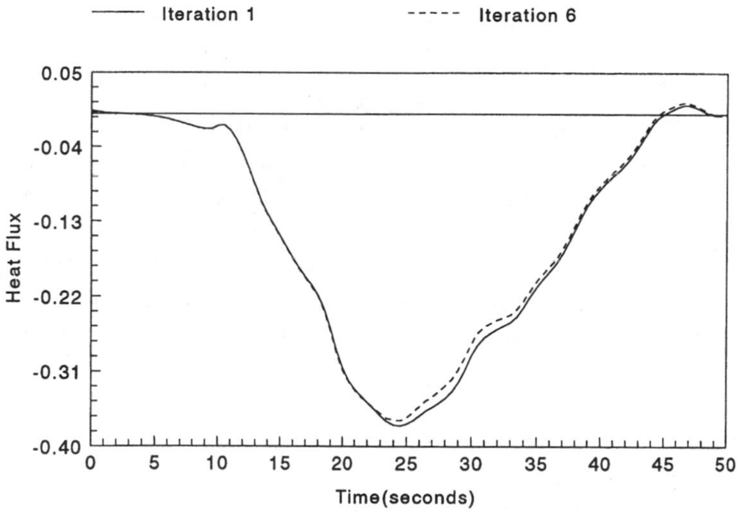

FIGURE 7.4
Heat flux 2, two-dimensional example.

model. In this example the truncated data has introduced some oscillations into the estimated heat fluxes.

A second two-dimensional example using actual measurements can be found in Trujillo and Wallis (1989). In this example a steel disk is quenched in an oil bath, and it is desired to estimate

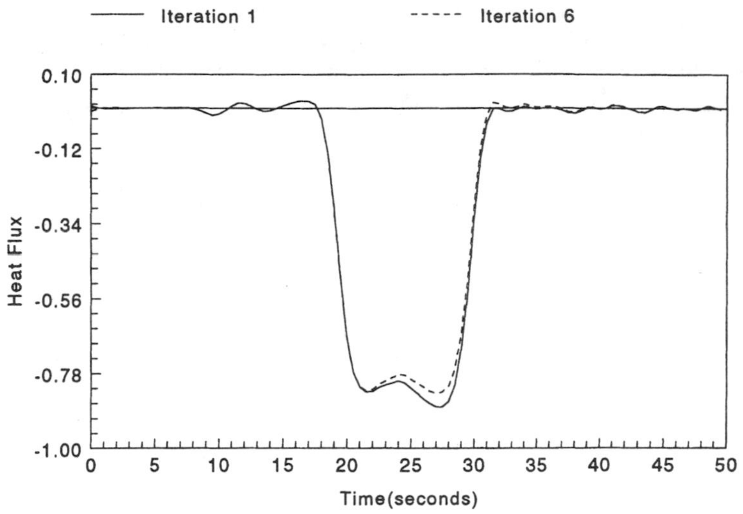

FIGURE 7.5
Heat flux 3, two-dimensional example.

the temperatures throughout the disk and the rate of heat being transferred from the disk during the quenching. The disk is heated to 2200°F and the oil is at 80°F. The data in this case come from 10 thermocouples embedded in the disk. Figure 7.6 shows an axisymmetric finite element model of the disk and the locations of the thermocouples. In this model x_j represents the 400 nodal temperatures. Seven unknown heat fluxes $\mathbf{q}_j$ were distributed around the disk. The model was constructed with more elements near the surfaces where the highest thermal gradients will occur during the quenching. The computer program INTEMP (1988), which is based on the techniques outlined in this chapter, was used to optimally estimate the heat fluxes. Figure 7.7 shows some typical results of the heat flux histories, two on the top and two on the bottom. It should be noted that in addition to matching the thermocouple data the model estimates the temperatures of the entire model. Of particular interest are the temperature histories of the surfaces in that they can be used to predict the metallurgical properties resulting from the quenching.

This example is nonlinear because the thermal properties of the steel vary with temperature. Currently, generalized cross validation has not been coupled with large nonlinear models

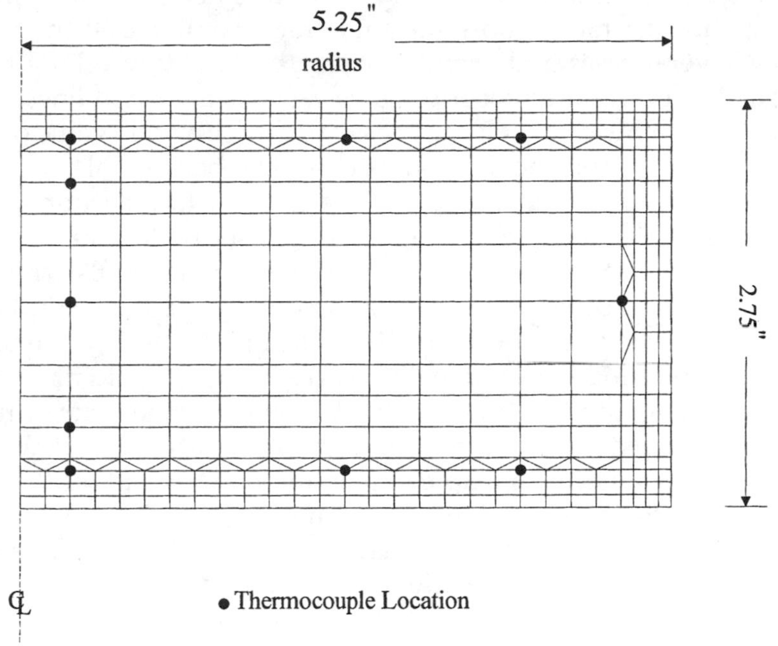

FIGURE 7.6
Axisymmetric finite element model showing positions of temperature measurement.

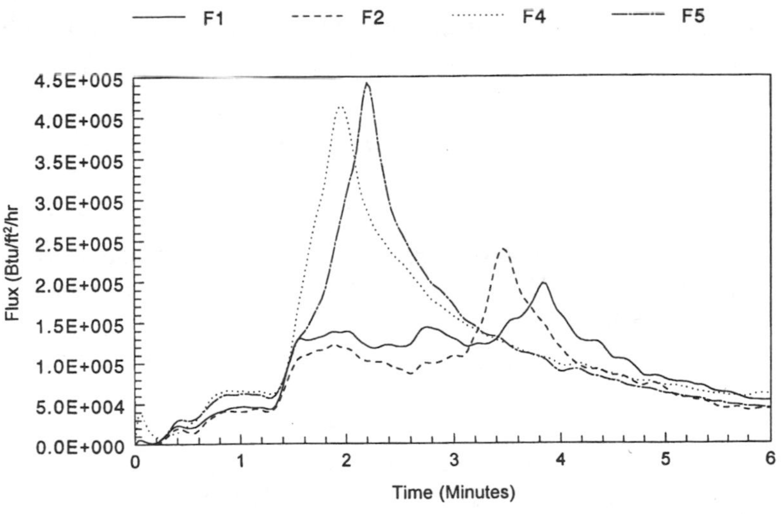

FIGURE 7.7
Variations of heat flux with time for the center area of disk.

because of excessive computational requirements. In cases like these the user may choose the smoothing parameter b by running several cases and examining the results. By weighing the matching of the thermocouple data against the smoothness of the heat flux histories, it is fairly easy to choose a good value of b.

Another promising method for choosing the optimal smoothing parameter is with the use of the L-curve (see Section 4.5). For this particular quenching experiment (Trujillo and Busby, 1994) the L-curve was constructed by solving the regularization problem using several values of the smoothing parameter b. The results were shown in Figure 4.19, which plots the F_{norm} versus the E_{norm} on a log-linear scale. For this case the E_{norm} did not vary much in magnitude. However, the L-shape characteristic of the curve is indeed present. Since this case represents real data there is no "true" answer to help evaluate the performance of the L-curve, but the corner value corresponds very well to the value of the smoothing parameter that was chosen by Trujillo and Wallis (1989) based on experience and intuition.

7.4 NONLINEAR SPRING EXAMPLE

This example is taken from Bellman et al. (1966) where it was used in conjunction with a parameter estimation problem. The model is a nonlinear spring, mass, and dashpot with an added nonlinear spring.

$$\ddot{x} + 3\dot{x} + 2x + 0.5x^3 = g(t) \tag{7.4-1}$$

The initial conditions are $x(0) = 2.0$ and $\dot{x}(0) = 0.0$. The inverse problem we are interested in solving is one where measurements have been taken on the displacement x and we wish to estimate the unknown forcing term g(t).

The above equation can be represented in vector-matrix form as

$$\begin{bmatrix} \dot{x} \\ \dot{v} \end{bmatrix} = \begin{bmatrix} 0 & 1 \\ -2 & -3 \end{bmatrix} \begin{bmatrix} x \\ v \end{bmatrix} + \begin{bmatrix} 0 \\ -0.5x^3 \end{bmatrix} + \begin{bmatrix} 0 \\ 1 \end{bmatrix} g(t) \tag{7.4-2}$$

As outlined in Section 7.2 the nonlinear term $0.5x^3$ can be approximated with

$$0.5x^3 \approx 0.5x^{*3} + 1.5x^{*2}\,(x - x^*) \tag{7.4-3}$$

where x* is the solution to a previous inverse problem. In order to investigate the practicability of using a constant Jacobian matrix, the term $1.5x^{*2}$ will be evaluated using a constant value of x* = 2.0. Using this approximation in Equation 7.4-2 gives

$$\begin{bmatrix} \dot{x} \\ \dot{v} \end{bmatrix} = \begin{bmatrix} 0 & 1 \\ -8 & -3 \end{bmatrix} \begin{bmatrix} x \\ v \end{bmatrix} + \begin{bmatrix} 0 \\ -0.5x^{*3} + 6x^* \end{bmatrix} + \begin{bmatrix} 0 \\ 1 \end{bmatrix} g(t) \tag{7.4-4}$$

This is the model that will be used for the nonlinear inverse problem. The unknown function is chosen as 5sin(t), which is used in a direct integration to generate simulated measurement of the displacement. The measurements were corrupted with a normally distributed noise with $\sigma = 0.10$. These data are shown in Figure 7.8, which also shows the estimated displacement after

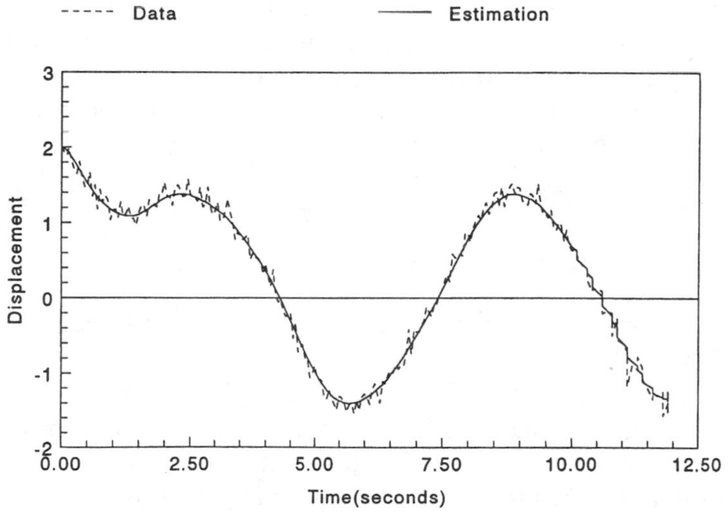

FIGURE 7.8
Comparison of data and estimated displacement.

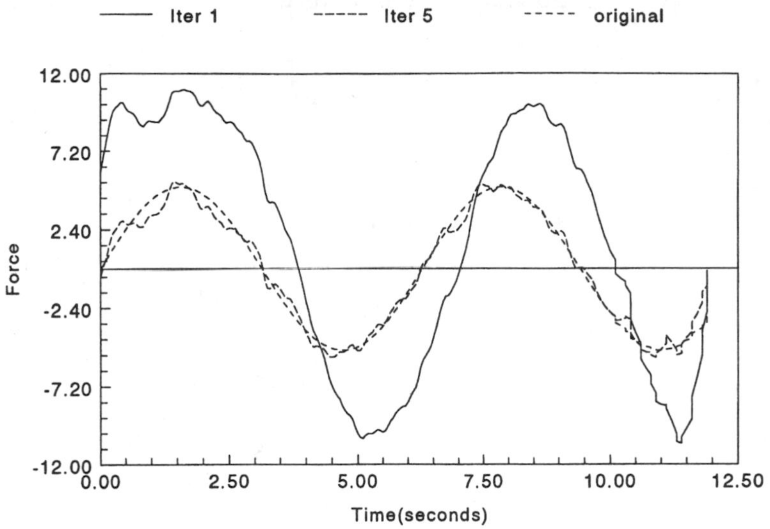

FIGURE 7.9
Comparison of forcing term.

five iterations. All of the inverse solutions were obtained using a smoothing parameter b = 1.E–3. The first iteration was obtained by setting the nonlinear terms in Equation 7.4-4 to zero. This results in crude estimation of the unknown force as shown in Figure 7.9. However, after the fifth iteration the estimations have converged as shown in the figure. The original function 5sin(t) is shown in the figure to illustrate that the nonlinear inverse methods produce a reasonable estimation.

7.5 SUCCESSIVE APPROXIMATION IN POLICY SPACE

Another technique one can use for the nonlinear inverse problem is the method of successive approximations in policy space (see Bellman and Dreyfus, 1962). This technique is based upon gradients in which the approach to an optimal policy is by successive steps. The method of derivation employs the familiar concepts and techniques of dynamic programming. We essentially guess a presumably nonoptimal decision sequence $\mathbf{z}_k$. By simple reasoning we derive a set of recurrence relations that can be used to evaluate the effect of a small change in the decision sequence. We then use this information about the effect of decision changes to generate a new, improved sequence of decisions.

The effect of changes in the new sequence is then evaluated. This iterative process is continued until no further improvement is possible. Each successive solution we obtain will be feasible for the problem, but not optimal.

7.5.1 Formulation

Consider a system of nonlinear difference equations

$$\mathbf{x}_{j+1} = \mathbf{g}_j(\mathbf{x}_j, \mathbf{z}_j) \qquad j = 1, 2, \ldots, N \tag{7.5.1-1}$$

where $\mathbf{x}$ is an $m \times 1$ vector representing the state variables, and $\mathbf{z}$ is a $k \times 1$ vector representing the forcing terms which are to be chosen optimally. The error criterion will be least squares given by the expression

$$E_N = \sum_{j=1}^{N} (\mathbf{Q}\mathbf{x}_j - \mathbf{d}_j, \mathbf{Q}\mathbf{x}_j - \mathbf{d}_j) + b(\mathbf{z}_j, \mathbf{z}_j) \tag{7.5.1-2}$$

where $\mathbf{x}_j$ is a vector at time jh and $\mathbf{Q}$ is a matrix which relates the state variables to the measurements $\mathbf{d}_j$, b is weighting for the regularization term $(\mathbf{z}_j, \mathbf{z}_j)$. Following Bellman (1957), define

$$f_n(\mathbf{x}_n) = \text{The value of E starting at state } \mathbf{x}_n \text{ and using the guessed policy } \mathbf{z}_j \tag{7.5.1-3}$$

Applying dynamic programming gives

$$f_n(\mathbf{x}_n) = (\mathbf{Q}\mathbf{x}_n - \mathbf{d}_n, \mathbf{Q}\mathbf{x}_n - \mathbf{d}_n) + b(\mathbf{z}_n, \mathbf{z}_n) + f_{n+1}(\mathbf{x}_{n+1}) \tag{7.5.1-4}$$

To determine the first-order effect of a change in the forcing terms $\mathbf{z}$ at time t(n), one needs to evaluate $\partial f_n / \partial \mathbf{z}$. By partial differentiation of Equation 7.5.1-4 with respect to $\mathbf{z}$ we have

$$\left[\frac{\partial f_n}{\partial \mathbf{z}}\right] = 2b\mathbf{z} + \left[\frac{\partial \mathbf{g}_n}{\partial \mathbf{z}}\right]^T \left[\frac{\partial f_{n+1}}{\partial \mathbf{x}}\right] \tag{7.5.1-5}$$

and also that

$$\left[\frac{\partial f_n}{\partial \mathbf{x}}\right] = 2\mathbf{Q}^T\left(\mathbf{Q}\mathbf{x}_n - \mathbf{d}_n\right) + \left[\frac{\partial g_n}{\partial \mathbf{x}}\right]^T \left[\frac{\partial f_{n+1}}{\partial \mathbf{x}}\right] \qquad (7.5.1\text{-}6)$$

The iterations for each timestep are

$$\mathbf{z}_n(\text{new}) = \mathbf{z}_n(\text{old}) + \delta \mathbf{z}_n$$

Recall that for each stage we have

$$\Delta f_n = \frac{\partial f_n}{\partial \mathbf{z}} \Delta \mathbf{z}$$

then choose $\Delta \mathbf{z}$ proportional to the gradient

$$\Delta \mathbf{z} = k \frac{\partial f_n}{\partial \mathbf{z}} \qquad (7.5.1\text{-}7)$$

$$\Delta f_n = k \left(\frac{\partial f_n}{\partial \mathbf{z}}\right)^2 \qquad (7.5.1\text{-}8)$$

k is determined from Equation 7.5.1-8 by setting Δf_n to be an incremental decrease in the current f_n. This in turn determines $\Delta \mathbf{z}$. The initial conditions are applied at the end point N.

$$\left[\frac{\partial f_N}{\partial \mathbf{x}}\right] = 2\mathbf{Q}^T\left(\mathbf{Q}\mathbf{x}_N - \mathbf{d}_N\right) \qquad (7.5.1\text{-}9)$$

Thus to obtain a solution we start with some reasonable guess as to the forcing terms. Using these forcing terms, Equation 7.5.1-1 can be integrated with the known initial conditions and the resulting **x**'s can be stored. Now Equations 7.5.1-5 and 7.5.1-6 can be integrated *backward* using the initial conditions, Equation 7.5.1-9. At each step we compute k and $\Delta \mathbf{z}$ from Equations 7.5.1-7 and 7.5.1-8. With a new forcing term we repeat the process and check for convergence. We continue until $\Delta \mathbf{z}$ is less than some specified small value.

7.5.2 Application to Nonlinear Systems

Let the differential equation be of the form

$$\dot{\mathbf{x}} + \mathbf{K}\mathbf{x} + \mathbf{n}(\mathbf{x}) = \mathbf{T}\mathbf{z}(t) \qquad (7.5.2\text{-}1)$$

where all the linear terms have been grouped into $\mathbf{Kx}$ leaving the nonlinear terms in $\mathbf{n}(\mathbf{x})$. Expanding $\mathbf{n}(\mathbf{x})$ with a Taylor's series expansion using the method outlined in Section 7.2 about some state $\mathbf{x}_0$ and letting $\mathbf{x}_0 = \mathbf{x}_i$ yields

$$\mathbf{x}_{i+1} = \mathbf{M}_i\mathbf{x}_i + \mathbf{s}_i + \mathbf{P}_i\mathbf{z}_i \qquad (7.5.2\text{-}2)$$

where

$$\begin{aligned}
\mathbf{M}_i &= \left[\mathbf{I} + (\mathbf{K} + \mathbf{A}_i)h/2\right]^{-1}\left[\mathbf{I} - (\mathbf{K} + \mathbf{A}_i)h/2\right] \\
\mathbf{s}_i &= -\left[\mathbf{I} + (\mathbf{K} + \mathbf{A}_i)h/2\right]^{-1} h\left(\mathbf{n}(\mathbf{x}_i) - \mathbf{A}_i\mathbf{x}_i\right) \qquad (7.5.2\text{-}3) \\
\mathbf{P}_i &= \left[\mathbf{I} + (\mathbf{K} + \mathbf{A}_i)h/2\right]^{-1}\mathbf{T}\mathbf{z}_i
\end{aligned}$$

It is important that these integration formulas remain stable. It should be noted that as $h \Rightarrow 0$, Equation 7.5.2-2 reduces to Equation 7.5.2-1, and $\mathbf{A}_i$ is not involved. Here $\mathbf{A}_i$ will be evaluated as the Jacobian matrix of $\mathbf{n}(\mathbf{x})$. With these definitions of $\mathbf{M}_i$, $\mathbf{s}_i$, and $\mathbf{P}_i$ the previous equations based on successive approximation can be used. For example, since

$$\mathbf{g}_i(\mathbf{x}_i, \mathbf{z}_i) = \mathbf{M}_i\mathbf{x}_i + \mathbf{s}_i + \mathbf{P}_i\mathbf{z}_i$$

then,

$$\frac{\partial \mathbf{g}_i}{\partial \mathbf{z}} = \mathbf{P}_i^T, \qquad \frac{\partial \mathbf{g}_i}{\partial \mathbf{x}} = \mathbf{M}_i^T$$

$$\frac{\partial f_i}{\partial \mathbf{z}} = 2b\mathbf{z}_i + \mathbf{P}_i^T\frac{\partial f_{i+1}}{\partial \mathbf{x}}, \qquad \frac{\partial f_i}{\partial \mathbf{x}} = 2\mathbf{Q}^T(\mathbf{Q}\mathbf{x}_i - \mathbf{d}_i) + \mathbf{M}^T\frac{\partial f_{i+1}}{\partial \mathbf{x}}$$

7.5.3 Nonlinear Spring Example

As an example of successive approximation in policy space, consider the nonlinear spring example given in Section 7.4 with the same initial conditions $x(0) = 2.0$, and $\dot{x}(0) = 0.0$ as before. Equation 7.4-2 can be solved as outlined above. The unknown function is again chosen to be 5sin(t). The measurements were corrupted with a normally distributed noise with $\sigma = 0.10$. These data are shown in Figure 7.10, which also shows the estimated

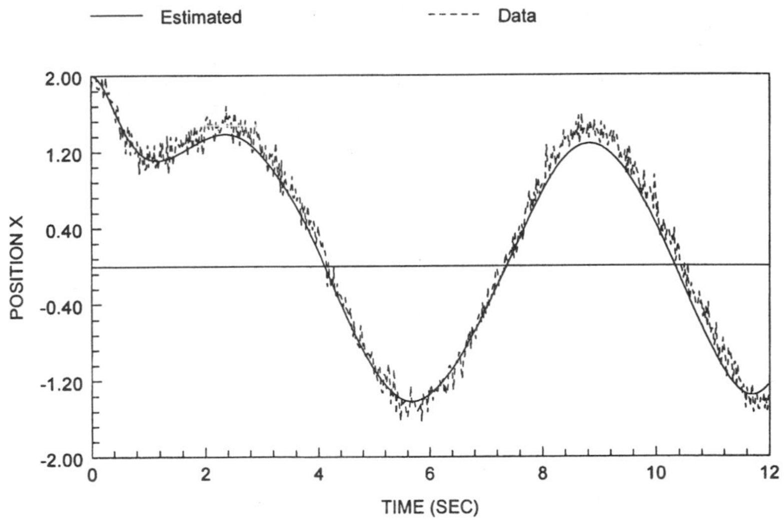

FIGURE 7.10
Comparison of data and estimated displacement.

displacement after 75 iterations, and Figure 7.11, which compares the forcing term. The original function 5sin(t) is shown in the figure to illustrate that the nonlinear inverse method produces a reasonable estimate. All of the inverse solutions were obtained using a smoothing parameter b = 5.E–3. Figure 7.12 demonstrates how the error is reduced after each iteration. However, for some problems the convergence rate may be slow and thus may require the selection and adjustment of several convergence parameters.

One method for smoothing the unknown forcing function is that of first-order regularization. The idea is to regulate the first derivative of the forcing function instead of the forces them-

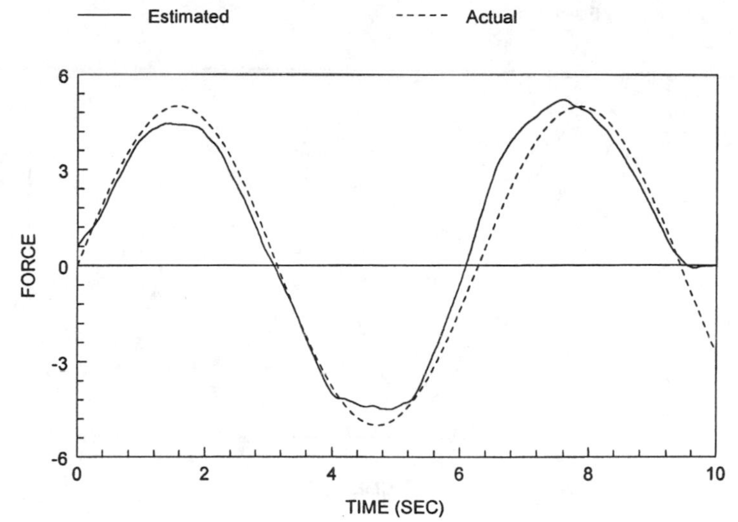

FIGURE 7.11
Comparison of forcing term (75 iterations).

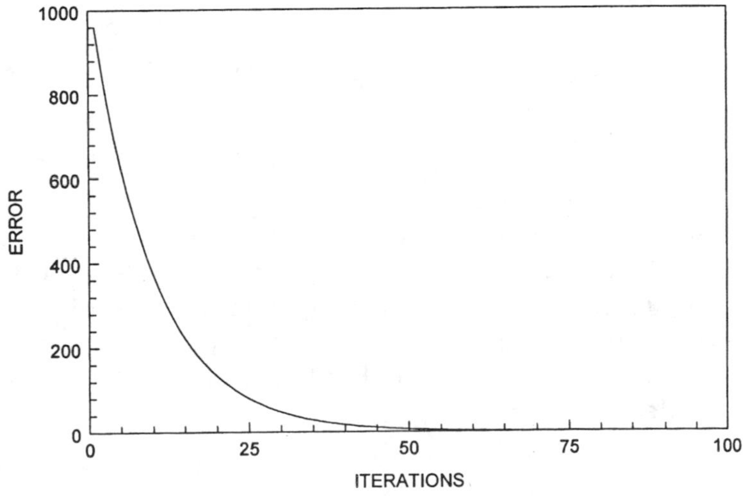

FIGURE 7.12
Least squares error versus number of iterations.

selves. This is done by adjoining the following differential equation to the dynamic model

$$\dot{g} = r \tag{7.5.3-1}$$

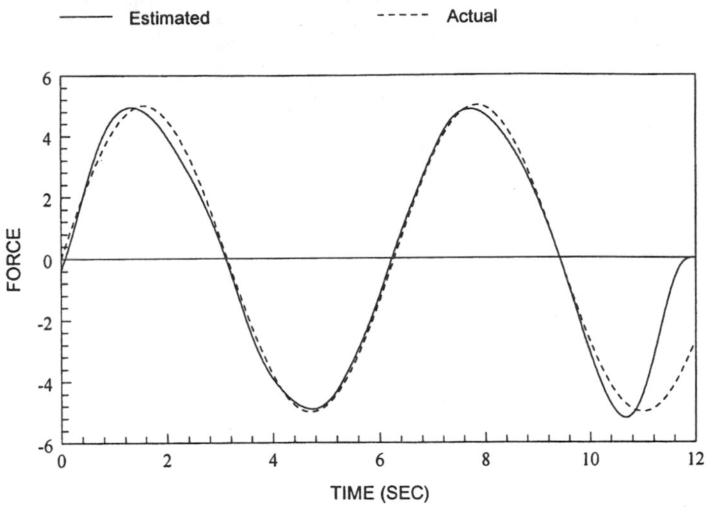

FIGURE 7.13
Comparison of forcing term using first-order regularization.

The new system becomes

$$\begin{bmatrix} \dot{x} \\ \dot{v} \\ \dot{g} \end{bmatrix} = \begin{bmatrix} 0 & -1 & 0 \\ 2 & 3 & -1 \\ 0 & 0 & 0 \end{bmatrix} \begin{bmatrix} x \\ v \\ g \end{bmatrix} + \begin{bmatrix} 0 \\ 0.5x^3 \\ 0 \end{bmatrix} + \begin{bmatrix} 0 \\ 0 \\ r \end{bmatrix} \tag{7.5.3-2}$$

and the error expression now uses r in the regulating term. Figure 7.13 shows the comparison of forcing terms using first-order regularization. The results show a reasonable estimation. Using first-order regularization however, requires more iterations for the error to converge. For this example 250 iterations were needed.

Chapter 8

Sequential Estimation and System Identification

8.1 INTRODUCTION

The systems identification problem or the parameter estimation problem arouses interest in all areas of engineering. The problem is to determine some or all of the system parameters based on measurements of the systems response. For example, a simple spring-mass-damper model with a base excitation is a reasonable representation of a skin-mounted accelerometer, which can be used to determine the motion of the underlying bone. It is clear that the layer of soft tissue covering the bone will influence the measurements as will the mass of the accelerometer itself. Thus, giving the spring-mass-damper model an initial displacement or velocity and measuring the response, one could then use this measured response to determine the effective spring constant and the effective damping constant of the tissue. This approach also has application to astronomy when one wishes to update the estimates of position and velocity as additional measurements or observations become available.

In this chapter we will present a simple approach to this problem based on least squares estimation. The objective is to arrive at a practical set of formulas that can be applied to these types of problems. In order to introduce systems identification or sequential estimation we will first discuss the simplest estimation of a state variable for the scalar case. The extension to the more general vector-matrix case will then be made. It will turn out that the system identification problem will become a nonlinear sequential estimation problem. This subject is related to stochastic models and Kalman filters. A more comprehensive viewpoint on this subject can be found in Astrom and Eykhoff (1971), Sage and Melsa (1971), and Mendel (1970).

8.2 SEQUENTIAL ESTIMATION

8.2.1 Simplest Case

The least squares sequential filter or estimator consists of two ideas. The first is fairly simple, while the second requires a dynamic programming formulation. First consider the calculation of a simple average of N data points d_N. The average x can be considered the minimum of

$$E_N = \sum_{i=1}^{N} (d_i - x)^2 \tag{8.2.1-1}$$

The average x is then

$$x = (1/N) \sum_{i=1}^{N} d_i \tag{8.2.1-2}$$

which involves N additions. Now suppose one were given an additional data point d_{N+1}. The new average could be recalculated using the above formula for N + 1 points. If one were repeatedly given additional data points, it would be inefficient to keep adding all of the points at each step. It is simple to show that if x_N is the average with N data points then the average x_{N+1} can be calculated with the following sequential formula

$$x_{N+1} = x_N + (d_{N+1} - x_N)/(N+1) \tag{8.2.1-3}$$

which involves a minimum of computations. Equation 8.2.1-3 can be thought of as an optimal filter in that it consists of a prediction for x_{N+1} plus a correction term based on the new measurement d_{N+1}. If the new measurement should happen to be equal to the predicted x_N, then no correction is made. Also, notice that for this filter the effect of the correction term becomes less as the number of data points becomes large. Another observation of this filter, related to dynamic programming, is that this filter can be viewed as successive optimal solutions to the problem. That is, one optimal solution x_{N+1} depends on a previously calculated optimal solution x_N. It is obvious that filters of this form would be very useful in real-time problems, especially for the vector matrix case.

Let us next expand our model to consider the case where the average may be varying. The case above considered x to be an absolute constant. It is easy to imagine a situation where the average was one value for a length of time and then changed to another value. If one wanted to estimate the current value of the average in real time instead of an overall average, a different model would be needed. One such model that allows the estimate of the average to vary is to minimize

$$E_N = \sum_{i=1}^{N} (d_i - x_i)^2 + k \sum_{i=2}^{N} (x_i - x_{i-1})^2 \tag{8.2.1-4}$$

This is now a much more difficult problem than before since the minimization is performed over all of the x_i's. That is, there are now N unknowns whereas before there was only one. As usual with these types of problems a weighting factor k has been introduced. This factor is used to control the dynamic response of the filter which will be discussed later.

Let us now consider the situation where the optimal x_i's have been calculated for N data points. A direct calculation would involve solving N linear equations for the N unknown x_i's. Denote the optimal estimate of the Nth point as x_N^*. Now suppose another data point d_{N+1} is added to the set. Another direct calculation involving N + 1 variables would yield an *entirely* new set of x_i's. Denote the optimal estimate of the N + 1 point as x_{N+1}^*. Since we are only interested in the last estimate x_{N+1}^*, is there a way to avoid unnecessary calculations as in the simple case discussed above? The answer is yes and a general derivation will be presented in the next section. These sequential filter formulas now require an additional variable r_N and are

$$r_{N+1} = 1 + kr_N/(k + r_N) \tag{8.2.1-5}$$

$$x_{N+1}^* = x_N^* + \left(d_{N+1} - x_N^*\right)/r_{N+1} \qquad (8.2.1\text{-}6)$$

The optimal estimates which lend themselves to real-time situations can now be easily updated with a minimum of computational effort.

The effect of the weighting term can now be seen. If k is set to a very small value, the error term $(x_i - x_{i-1})^2$ is not important, and this will force the filter to estimate $x_{N+1}^* = d_{N+1}$, i.e., use the latest measurement (r_{N+1} will be approximately 1.0). On the other hand, if k is set to a very large value, the error term will force the x_N^*'s to remain nearly constant and the simple average will be calculated, i.e., r_{N+1} will be approximately equal to N + 1. The choice of k usually requires some experimentation with actual or simulated data.

The above examples are fairly simple and lend themselves to structured analytical solutions. More realistic problems will require numerical or computational solutions. For example, consider a slight variation of the simple average problem: suppose there exist *two* averages in the data being sampled. That is, each data point could apply to one of two averages and it is desired to optimally estimate these averages simultaneously. The error equation is now

$$E_N(c_1, c_2) = \sum_{i=1}^{N} (d_i - c_1)^2 + \sum_{i=1}^{N} (d_i - c_2)^2$$

Now this expression could easily be minimized with respect to the two averages c_1 and c_2 if the data were properly segregated into the correct group. In other words, each time one is given a new data point it must first be decided which average it is associated with — c_1 or c_2. However, the averages are unknown. In this case, using the concept of dynamic programming, one can define a functional $F_N(c_1,c_2)$ as:

$F_N(c_1,c_2)$ = the value of E_N (c_1,c_2) distributing the data points in an optimal manner.

Since the functional depends on any state (c_1,c_2) it is easy to rationalize that the data point should be associated with the closest state. Mathematically, this is expressed as

$$F_{N+1}(c_1, c_2) = \min\begin{cases}(d_{N+1} - c_1)^2 + F_N(c_1, c_2)\\(d_{N+1} - c_2)^2 + F_N(c_1, c_2)\end{cases}$$

As before, this functional represents the running summations for all states, and the optimal estimate of the two averages at any stage N is obtained by minimizing $F_N(c_1,c_2)$ over c_1 and c_2. Due to the complexity of this problem, it is not possible to derive recurrence formulas that only involve the optimal estimates.

Exercise

1. Derive Equation 8.2.1-3.
2. Derive the N linear equations required to solve for the optimal x_i's of Equation 8.2.1-4. Is there a special structure that lends itself to a simple computational solution?
3. Derive the N + 1 linear equations for N + 1 points and verify Equations 8.2.1-5 and 8.2.1-6.

8.2.2 General Scalar Case

Consider a dynamic system governed by the expression

$$x_{i+1} = a_i x_i + g_i + \varepsilon_i \tag{8.2.2-1}$$

and suppose that a measurement d_i is taken at every step and related to x_i by

$$d_i = h_i x_i + \eta_i \tag{8.2.2-2}$$

where a_i and h_i are known and g_i is a known forcing term. In addition, ε_i and η_i represent noise with the usual properties. It is desired to optimally estimate x_i using a least squares criterion. In addition, suppose that N measurements have been taken and that all of the x_i's have been optimally estimated. Now another measurement d_{N+1} is taken. We would like to optimally estimate only the current x_{N+1} without having to recalculate all the previous optimal estimates. This is necessary in a real-time constraint where only the current state is important, and it is also very useful in estimating nonlinear systems when a linearization approximation requires an estimate of the state variables.

Since neither the dynamical equation nor the observation was exact, we form an estimation criterion function which pays attention to both dynamical and observational sources of error. The least square criterion function for N measurements is given as

$$E_N = \sum_{i=1}^{N} (d_i - h_i x_i)^2 + k \sum_{i=2}^{N} (x_i - a_{i-1} x_{i-1} - g_{i-1})^2 \tag{8.2.2-3}$$

Consider the second term of Equation 8.2.2-3. It is usually included because of the noise driving force ε_i (see Equation 8.2.2-1). Another reason for including this term is that in these types of problems we will only have an estimate of x_{i-1} and it does not make much sense to set the next estimate x_i exactly equal to $a_{i-1}x_{i-1} + g_{i-1}$. Of course the weighting parameter k can be used to control the filter's behavior by balancing the measurement against satisfying the dynamical equations. With some experimentation one can see the effects of varying the weighting parameter.

To restate the problem, we wish to minimize Equation 8.2.2-3 with respect to all x_i, i.e., $x_1, x_2, \ldots, x_N$. Then suppose we add another data point d_{N+1}, which gives

$$E_{N+1} = \sum_{i=1}^{N+1} (d_i - h_i x_i)^2 + k \sum_{i=1}^{N+1} (x_i - a_{i-1} x_{i-1} - g_{i-1})^2 \tag{8.2.2-4}$$

We now wish to minimize E_{N+1} with respect to $x_1, x_2, \ldots, x_{N+1}$. It is very important to note that *all* of the minimizing x_i's will change by adding the one data point. In addition, we only want the estimate of x_{N+1} (the current state) in order for the computations to be kept to a minimum. This is another case where dynamic programming can be used to provide the sequential formulas.

To obtain a solution, first define $\Theta_N(c)$ as the minimum of E_N conditioned on $x_N = c$, where c is arbitrary. That is, suppose x_N is fixed at c and we minimize E_N with respect to $x_1, x_2, \ldots, x_{N-1}$. The value of E_N is $\Theta_N(c)$. Using a dynamic programming argument, it follows that

$$\Theta_{N+1}(c) = \min_{x_N}\Big[(d_{N+1} - h_{N+1}c)^2 + k(c - a_N x_N - g_N)^2 + \Theta_N(x_N)\Big] \tag{8.2.2-5}$$

This equation will give the function $\Theta_{N+1}(c)$ which is valid for all $x_{N+1} = c$. However, the optimal estimate of x_{N+1} is the one that *minimizes* $\Theta_{N+1}(c)$ with respect to c. Hence, we have separated the problem into two simple parts.

A recursive formula for Θ_N can be obtained by first considering N = 1. At the beginning only one data point is available which is d_1, thus

$$\Theta_1(c) = (d_1 - h_1 c)^2 \tag{8.2.2-6}$$

and there are no further terms involved. Now we consider the next step where

$$\Theta_2(c) = \min_{x_1}\Big[(d_2 - h_2 c)^2 + k(c - a_1 x_1 - g_1)^2 + \Theta_1(x_1)\Big] \tag{8.2.2-7}$$

It is easy to show that $\Theta_N(c)$ will be of the form

$$\Theta_N(c) = r_N c^2 - 2 s_N c + q_N \tag{8.2.2-8}$$

and recursive formulas for r_N, s_N, and q_N can be obtained. Substituting Equation 8.2.2-8 into Equation 8.2.2-5 yields

$$r_{N+1}c^2 - 2s_{N+1}c + q_{N+1} = \min_{x_N}\Big[(d_{N+1} - h_{N+1}c)^2 + k(c - a_N x_N - g_N)^2 + r_N x_N^2 - 2 s_N x_N + q_N\Big] \tag{8.2.2-9}$$

Minimizing Equation 8.2.2-9 with respect to x_N yields

$$x_N^* = (r_N + k a_N^2)^{-1}\big[s_N + k a_N (c - g_N)\big] \tag{8.2.2-10}$$

Substituting Equation 8.2.2-10 into Equation 8.2.2-9 and equating like coefficients in c gives

$$r_{N+1} = h_{N+1}^2 + kr_N / \left(r_N + ka_N^2\right) \tag{8.2.2-11}$$

$$s_{N+1} = d_{N+1}h_{N+1} + ka_N s_N m_N^{-1} - m_N^{-1}k^2 a_N^2 g_N + kg_N \tag{8.2.2-12}$$

$$q_{N+1} = d_{N+1}^2 - m_N^{-1}\left(s_N - ka_N g_N\right)^2 + g_N^2 + q_N \tag{8.2.2-13}$$

where $m_N = r_N + ka_N^2$. The initial conditions for r_N, s_N, and q_N are obtained from Equation 8.2.2-6 and are

$$r_1 = h_1^2, \quad s_1 = h_1, \quad q_1 = d_1^2 \tag{8.2.2-14}$$

The optimal estimate of x_N is obtained by minimizing $\Theta_N(c)$, which we denote by c^*. Hence we find

$$c_N^* = s_N / r_N \tag{8.2.2-15}$$

The next optimal estimate would be given by

$$c_{N+1}^* = s_{N+1} / r_{N+1} \tag{8.2.2-16}$$

It is now possible to derive formulas involving only the current optimal estimates c_N^* and c_{N+1}^*. The final results will now involve two recurrence formulas which are given as Equation 8.2.2-11 for r_{N+1} and

$$\begin{aligned} c_{N+1}^* &= a_N c_N^* + g_N \\ &+ \left(h_{N+1}/r_{N+1}\right)\left[d_{N+1} - h_{N+1}\left(a_N c_N^* + g_N\right)\right] \end{aligned} \tag{8.2.2-17}$$

These recurrence formulas are the deterministic versions of the Kalman filter, which is based on stochastic models. The form of the optimal filter consists of a prediction plus a correction term based on the new measurement d_{N+1}.

Exercise

1. Carry out the minimization indicated by Equation 8.2.2-7 and obtain the final formula for $\Theta_2(c)$.
2. Derive Equation 8.2.2-17.
3. The fact that all of the optimal x_i's are changed each time another data point is added leads to cumbersome notation in the literature. How would one compute all of these x_i's using Equation 8.2.2-10? These are called the smoothed estimates.

8.2.3 Example

As an example, consider the simple first-order model

$$dx/dt + bx = 0 \tag{8.2.3-1}$$

The problem is to estimate the state x from some measurements d_i taken at discrete times. Using the exponential technique described in Chapter 2, we can replace the above equation with an equivalent discrete system given by

$$x_{i+1} = e^{-b\Delta t} x_i \tag{8.2.3-2}$$

where Δt is the timestep. To apply our recursive formula for the optimal estimates we need to evaluate the initial values. Recall our dynamic system governed by the expressions

$$x_{i+1} = a_i x_i + g_i + \varepsilon_i$$

$$d_{i+1} = h_i x_i + \eta_i$$

Comparing these equations with Equation 8.2.3-1, we find that

$$a_i = e^{-b\Delta t}, \quad g_i = 0, \quad h_i = 1.0 \tag{8.2.3-3}$$

Also, from Equations 8.2.2-6 and 8.2.2-8 we have

$$\Theta_1(c) = r_1 c^2 - 2s_1 c + q_1 = (d_1 - c)^2$$

which implies that the initial conditions for the filter are

$$q_1 = d_1^2, \quad r_1 = 1.0, \quad \text{and} \quad s_1 = d_1$$

In addition, from Equation 8.2.2-14 we have

$$c_1^* = s_1/r_1$$

We find for the initial conditions that $c_1^* = d_1$. Now we can apply our formula involving only the optimal estimates

$$c_{N+1}^* = a_N c_N^* + \left(d_{N+1} - a_N c_N^*\right)/r_{N+1} \qquad (8.2.3\text{-}4)$$

Using t = 0.1, b = 1.0, and c = 10.0, Equation 8.2.3-4 was solved for the displacement x. A random noise level of $\sigma = 0.1$ was then added to the displacement. The resulting noisy data are shown in Figure 8.1 along with the optimal estimation. The weighting parameter k used was 50.0. As the weighting parameter k becomes smaller, the solution tends to follow the data as shown in Figure 8.2.

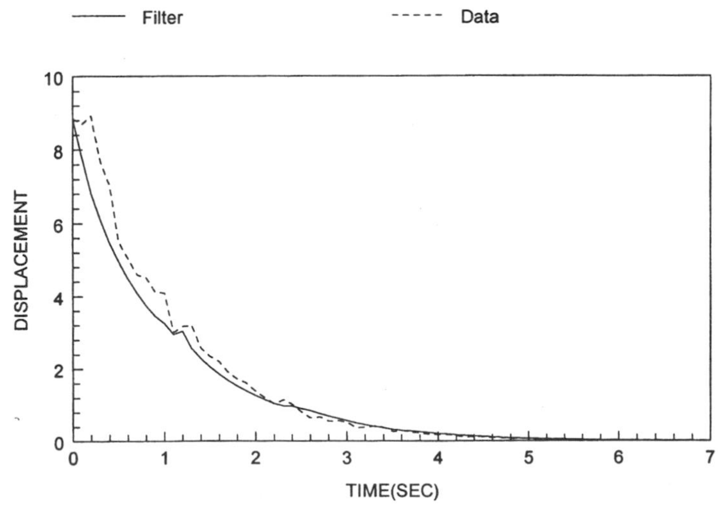

FIGURE 8.1
Comparison of estimated and original signal with noise.

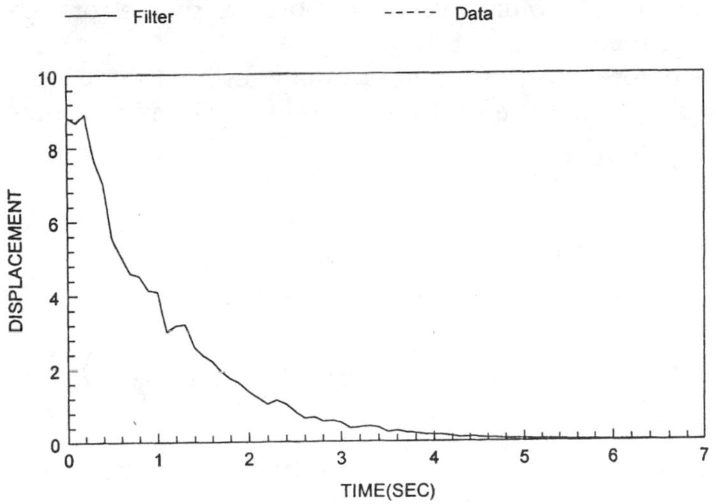

FIGURE 8.2
Comparison of estimated and original signal with k = 0.01.

8.3 MULTIDIMENSIONAL SEQUENTIAL ESTIMATION

The analogs of the equations developed for the scalar case will now be presented for the vector case. The development here closely follows that presented in Section 8.2.2. Consider a dynamical process described by an n dimensional system of difference equations

$$\mathbf{x}_{i+1} = \mathbf{M}_i \mathbf{x}_i + \mathbf{g}_i + \varepsilon_i \tag{8.3-1}$$

where $\mathbf{x}_i$ is an $(n \times 1)$ column vector of state variables, $\mathbf{M}_i$ is an $(n \times n)$ transfer matrix, $\mathbf{g}_i$ is an $(n \times 1)$ known forcing term and ε_i is an $(n \times 1)$ vector of dynamical errors. It is assumed that observations on the state vector are obtained at every step and related to $\mathbf{x}_i$ by

$$\mathbf{d}_i = \mathbf{H}_i \mathbf{x}_i + \eta_i \tag{8.3-2}$$

where $\mathbf{d}_i$ is an $(m \times 1)$ column vector of measurements, $\mathbf{H}_i$ is an $(m \times n)$ matrix relating the measurements to the state vector and

η_i is an (m × 1) column vector of observational errors. Usually m is less than n.

As before, it is desired to optimally estimate $\mathbf{x}_i$ using a least squares criterion. The least squares criterion for N measurements is

$$E_N = \sum_{i=1}^{N} \left(\mathbf{d}_i - \mathbf{H}_i\mathbf{x}_i, \mathbf{d}_i - \mathbf{H}_i\mathbf{x}_i\right) + \sum_{l=2}^{N} \left(\mathbf{x}_i - \mathbf{M}_{i-1}\mathbf{x}_{i-1} - \mathbf{g}_{i-1}, \mathbf{K}\left(\mathbf{x}_i - \mathbf{M}_{i-1}\mathbf{x}_{i-1} - \mathbf{g}_{i-1}\right)\right) \tag{8.3-3}$$

where (**x**,**y**) denotes the inner product of two vectors and **K** represents an n × n weighting matrix.

To restate the problem, we wish to minimize Equation 8.3-3 with respect to all $\mathbf{x}_i$, i.e., $(\mathbf{x}_1,\mathbf{x}_2,\ldots,\mathbf{x}_N)$. Then suppose we add another data point $\mathbf{d}_{N+1}$. We would then minimize over $(\mathbf{x}_1,\mathbf{x}_2,\ldots,\mathbf{x}_{N+1})$. In addition, we are only interested in the current estimate. We define, as before, $\Theta_N(\mathbf{c})$ to be the minimum of E_N conditioned on $\mathbf{x}_N = \mathbf{c}$. That is, suppose $\mathbf{x}_N$ is fixed at **c** and we minimize E_N with respect to $\mathbf{x}_1,\mathbf{x}_2,\ldots,\mathbf{x}_{N-1}$. The value of E_N is $\Theta_N(\mathbf{c})$. Using a dynamic programming argument it follows that

$$\Theta_{N+1}(\mathbf{c}) = \min_{\mathbf{x}_N}\Big[\left(\mathbf{d}_{N+1} - \mathbf{H}_{N+1}\mathbf{c}, \mathbf{d}_{N+1} - \mathbf{H}_{N+1}\mathbf{c}\right) + \left(\mathbf{c} - \mathbf{M}_N\mathbf{x}_N - \mathbf{g}_N, \mathbf{K}\left(\mathbf{c} - \mathbf{M}_N\mathbf{x}_N - \mathbf{g}_N\right)\right) + \Theta\left(\mathbf{x}_N\right)\Big] \tag{8.3-4}$$

Equation 8.3-4 will give the function $\Theta_{N+1}(\mathbf{c})$, which is valid for all $\mathbf{x}_{N+1} = \mathbf{c}$. However, the optimal estimate of $\mathbf{x}_{N+1}$ is the one that minimizes $\Theta_{N+1}(\mathbf{c})$ with respect to **c**. One can show that all $\Theta_N(\mathbf{c})$ will be of the form

$$\Theta_N(\mathbf{c}) = q_N - 2\left(\mathbf{c}, \mathbf{s}_N\right) + \left(\mathbf{c}, \mathbf{R}_N\mathbf{c}\right) \tag{8.3-5}$$

where q_N is a scalar, $\mathbf{s}_N$ is an (n × 1) recursive vector, and $\mathbf{R}_N$ is an (n × n) matrix. Recursive formulas for $\mathbf{s}_N$ and $\mathbf{R}_N$ can be obtained by substituting Equation 8.3-5 into Equation 8.3-4 which gives

$$
\begin{aligned}
q_{N+1} - 2(\mathbf{c}, \mathbf{s}_{N+1}) + (\mathbf{c}, \mathbf{R}_{N+1}\mathbf{c}) = \min_{\mathbf{x}_N}\big[&(\mathbf{d}_{N+1}, \mathbf{d}_{N+1}) \\
&- 2(\mathbf{d}_{N+1}, \mathbf{H}_{N+1}\mathbf{c}) + (\mathbf{H}_{N+1}\mathbf{c}, \mathbf{H}_{N+1}\mathbf{c}) + (\mathbf{c}, \mathbf{K}\mathbf{c}) \\
&- 2(\mathbf{c}, \mathbf{K}\mathbf{M}_N\mathbf{x}_N) - 2(\mathbf{c}, \mathbf{K}\mathbf{g}_N) + (\mathbf{M}_N\mathbf{x}_N, \mathbf{K}\mathbf{M}_N\mathbf{x}_N) \\
&+ 2(\mathbf{M}_N\mathbf{x}_N, \mathbf{K}\mathbf{g}_N) + (\mathbf{g}_N, \mathbf{K}\mathbf{g}_N) - 2(\mathbf{x}_N, \mathbf{s}_N) \\
&+ (\mathbf{x}_N, \mathbf{R}_N\mathbf{x}_N) + q_N\big]
\end{aligned}
\qquad (8.3\text{-}6)
$$

Minimizing Equation 8.3-6 with respect to $\mathbf{x}_N$ yields

$$\mathbf{x}_N^* = \mathbf{Q}_N\mathbf{M}_N^T\mathbf{K}(\mathbf{c} - \mathbf{g}_N) + \mathbf{Q}_N\mathbf{s}_N \qquad (8.3\text{-}7)$$

where

$$\mathbf{Q}_N^{-1} = \mathbf{M}_N^T\mathbf{K}\mathbf{M}_N + \mathbf{R}_N \qquad (8.3\text{-}8)$$

$\mathbf{Q}_N$ is an $(n \times n)$ matrix. Substituting Equation 8.3-7 into Equation 8.3-6 and equating like coefficients in $\mathbf{c}$ yields

$$\mathbf{s}_{N+1} = \mathbf{H}_{N+1}^T\mathbf{d}_{N+1} + \mathbf{K}\mathbf{g}_N - \mathbf{K}\mathbf{M}_N\mathbf{Q}_N\mathbf{M}_N^T\mathbf{K}\mathbf{g}_N + \mathbf{K}\mathbf{M}_N\mathbf{Q}_N\mathbf{s}_N \qquad (8.3\text{-}9)$$

$$\mathbf{R}_{N+1} = \mathbf{H}_{N+1}^T\mathbf{H}_{N+1} + \mathbf{K} - \mathbf{K}\mathbf{M}_N\mathbf{Q}_N\mathbf{M}_N^T\mathbf{K} \qquad (8.3\text{-}10)$$

This equation for $\mathbf{R}_{N+1}$ is a discrete version of the Riccati equation. The optimal estimate of $\mathbf{x}_N$ is produced by minimizing $\Theta_N(\mathbf{c})$ which is denoted by $\mathbf{c}^*$. Thus

$$2\mathbf{R}_N\mathbf{c}_N^* = 2\mathbf{s}_N = 0$$

or

$$\mathbf{c}_N^* = \mathbf{R}_N^{-1}\mathbf{s}_N \qquad (8.3\text{-}11)$$

Equation 8.3-11 can serve as the final calculation, or as we did for the scalar case, we can consider the calculation given by

$$\mathbf{c}^*_{N+1} = \mathbf{R}^{-1}_{N+1}\mathbf{s}_{N+1} \tag{8.3-12}$$

Using Equations 8.3-8, 9, and 10 we find

$$\mathbf{c}^*_{N+1} = \mathbf{M}_N\mathbf{c}^*_N + \mathbf{g}_N + \mathbf{R}^{-1}_{N+1}\mathbf{H}^T_{N+1}\left[\mathbf{d}_{N+1} - \mathbf{H}_{N+1}\left(\mathbf{M}_N\mathbf{c}^*_N + \mathbf{g}_N\right)\right] \tag{8.3-13}$$

which involves only the optimal estimates.

Since Equation 8.3-13 involves only the inverse of $\mathbf{R}_{N+1}$, it is convenient to obtain a recurrence formula that involves only the inverse. To achieve this, consider Equations 8.3-8 and 8.3-10. Applying the Sherman Morrison formula (see Chapter 2) to Equation 8.3-8 we find

$$\mathbf{Q}_N = \mathbf{R}^{-1}_N - \mathbf{R}^{-1}_N\mathbf{M}^T_N\left(\mathbf{K}^{-1} + \mathbf{M}_N\mathbf{R}^{-1}_N\mathbf{M}^T_N\right)^{-1}\mathbf{M}_N\mathbf{R}^{-1}_N \tag{8.3-14}$$

Considering the last term of Equation 8.3-10 and the results of Equation 8.3-14, we can obtain the resulting expression

$$\mathbf{R}_{N+1} = \mathbf{H}^T_{N+1}\mathbf{H}_{N+1} + \left(\mathbf{K}^{-1} + \mathbf{M}_N\mathbf{R}^{-1}_N\mathbf{M}^T_N\right)^{-1} \tag{8.3-15}$$

Define

$$\mathbf{G}_N = \mathbf{K}^{-1} + \mathbf{M}_N\mathbf{R}^{-1}_N\mathbf{M}^T_N \tag{8.3-16}$$

where $\mathbf{G}_N$ is an (n × n) matrix. Equation 8.3-15 can be written as

$$\mathbf{R}_{N+1} = \mathbf{H}^T_{N+1}\mathbf{H}_{N+1} + \mathbf{G}^{-1}_N \tag{8.3-17}$$

Applying the Sherman Morrison formula again to Equation 8.3-17 yields an expression with only one inverse. Thus, the final form of the sequential filter is given as $\mathbf{P} = \mathbf{R}^{-1}_N$.

$$\mathbf{G}_N = \mathbf{M}_N\mathbf{P}_N\mathbf{M}^T_N + \mathbf{K}^{-1} \tag{8.3-18}$$

$$\mathbf{P}_{N+1} = \mathbf{G}_N - \mathbf{G}_N\mathbf{H}^T_{N+1}\left[\mathbf{I} + \mathbf{H}_{N+1}\mathbf{G}_N\mathbf{H}^T_{N+1}\right]^{-1}\mathbf{H}_{N+1}\mathbf{G}_N \tag{8.3-19}$$

$$\mathbf{c}^*_{N+1} = \mathbf{M}_N\mathbf{c}^*_N + \mathbf{g}_N + \mathbf{P}_{N+1}\mathbf{H}^T_{N+1}\left[\mathbf{d}_{N+1} - \mathbf{H}_{N+1}\left(\mathbf{M}_N\mathbf{c}^*_N + \mathbf{g}_N\right)\right] \quad (8.3\text{-}20)$$

Equations 8.3-18, 19, and 20 are the deterministic equivalent of the Kalman filter for multidimensional systems. Notice that the only inverse in Equation 8.3-19 involves a matrix of dimension $m \times m$.

The initial conditions are arrived at by considering the first data point in the error expression

$$E_1 = \left(\mathbf{d}_1 - \mathbf{H}_1\mathbf{c}, \mathbf{d}_1 - \mathbf{H}_1\mathbf{c}\right)$$

Expanding this expression gives

$$E_1 = \left(\mathbf{d}_1, \mathbf{d}_1\right) - 2\left(\mathbf{H}_1^T\mathbf{d}_1, \mathbf{c}\right) + \left(\mathbf{H}_1^T\mathbf{H}_1\mathbf{c}, \mathbf{c}\right)$$

which can be compared to Equation 8.3-5 to give

$$q_1 = \left(\mathbf{d}_1, \mathbf{d}_1\right) \quad (8.3\text{-}21)$$

$$\mathbf{s}_1 = \mathbf{H}_1^T\mathbf{d}_1 \quad (8.3\text{-}22)$$

$$\mathbf{R}_1 = \mathbf{H}_1^T\mathbf{H}_1 \quad (8.3\text{-}23)$$

Since the optimal estimates involve the inverse of **R**, there is a question of whether that inverse exists, especially at the beginning of the process. This is natural since there must be a physical connection between the state vector x, its dimension, and the number of measurements. For example, it is not possible to estimate the displacement, velocity, and acceleration from only a single data point. From a practical viewpoint it is always possible to augment the error criteria with a term involving an initial estimate of the state vector. This will assure that the filter equations are always tractable. There can still be difficulties with the filter if part of the model or the measurements are physically decoupled from each other. These topics are beyond the scope of this presentation. See Bryson and Ho (1975) for discussions of observability. There has also been a considerable amount of investigation into the computational efficiency of these equations (see Bierman, 1977).

Exercise

1. Add a term involving an initial estimation of the state vector $\mathbf{z}_1$, $(\mathbf{x}_1 - \mathbf{z}_1, \mathbf{W}(\mathbf{x}_1 - \mathbf{z}_1))$ and rederive the filter equations. Is it necessary for the inverse of $\mathbf{W}$ to exist?
2. Is it possible to compute the matrices $\mathbf{R}_N$ without having to calculate its inverse? That is, can the filter be started after a sufficient number of data points have been taken to insure that the inverse of $\mathbf{R}_N$ exists?
3. Using the model of Section 6.2, write a computer program to estimate in real time the displacement, velocity, and acceleration from a displacement measurement.

8.3.1 Application To Nonlinear Systems

In the previous section, the linear difference equation was used to represent the system

$$\mathbf{x}_{i+1} = \mathbf{M}_i\mathbf{x}_i + \mathbf{g}_i \tag{8.3.1-1}$$

One way that this can be used to represent a continuous nonlinear system is the following: Let the original differential equation be of the form

$$\dot{\mathbf{x}} + \mathbf{K}\mathbf{x} + \mathbf{n}(\mathbf{x}) = \mathbf{f}(t) \tag{8.3.1-2}$$

where all of the linear terms have been grouped into $\mathbf{Kx}$ leaving the nonlinear terms in $\mathbf{n}(\mathbf{x})$. Expanding $\mathbf{n}(\mathbf{x})$ with a Taylor's expansion about some state $\mathbf{x}_0$ gives

$$\mathbf{n}(\mathbf{x}) = \mathbf{n}(\mathbf{x}_0) + \mathbf{A}(\mathbf{x} - \mathbf{x}_0) \tag{8.3.1-3}$$

where $\mathbf{A}$ represents the Jacobian matrix of $\mathbf{n}$ evaluated at $\mathbf{x}_0$. This gives

$$\dot{\mathbf{x}} + (\mathbf{K} + \mathbf{A})\mathbf{x} = -\mathbf{n}(\mathbf{x}_0) + \mathbf{A}\mathbf{x}_0 + \mathbf{f}(t) \tag{8.3.1-4}$$

Using the Crank–Nicolson method described in Chapter 2, Equation 8.3.1-4 can be put into discrete form as

$$\begin{aligned}\left[\mathbf{I}+(\mathbf{K}+\mathbf{A})h/2\right]\mathbf{x}_{i+1} &= \left[\mathbf{I}-(\mathbf{K}+\mathbf{A})h/2\right]\mathbf{x}_i \\ &\quad - h\mathbf{n}(\mathbf{x}_0) + h\mathbf{A}\mathbf{x}_0 + h\mathbf{f}\end{aligned} \qquad (8.3.1\text{-}5)$$

where h is the timestep. Now let $\mathbf{x}_0 = \mathbf{x}_i$, then the terms in Equation 8.3.1-1 become

$$\mathbf{M}_i = \left[\mathbf{I}+(\mathbf{K}+\mathbf{A}_i)h/2\right]^{-1}\left[\mathbf{I}-(\mathbf{K}+\mathbf{A}_i)h/2\right] \qquad (8.3.1\text{-}6)$$

$$\mathbf{g}_i = -\left[\mathbf{I}+(\mathbf{K}+\mathbf{A}_i)h/2\right]^{-1} h\left(\mathbf{n}(\mathbf{x}_i) - \mathbf{A}_i\mathbf{x}_i - \mathbf{f}_i\right) \qquad (8.3.1\text{-}7)$$

It is important that these integration formulas remain stable. It should be noted that as h approaches 0, Equation 8.3.1-5 with $\mathbf{x}_0 = \mathbf{x}_i$, reduces to Equation 8.3.1-2, and $\mathbf{A}_i$ is not involved. There is some evidence that $\mathbf{A}_i$ can be used to stabilize the integration process since it is somewhat arbitrary. However, in this section $\mathbf{A}_i$ will be evaluated as the Jacobian matrix of $\mathbf{n}(\mathbf{x})$. With these definitions of $\mathbf{M}_i$ and $\mathbf{g}_i$ the previous filter equations can now be used directly for a nonlinear equation. The Jacobian matrix and the nonlinear term can now be evaluated at the previous estimate of the state vector.

The above formulas are only one way in which to convert a continuous nonlinear multidimensional system to a quasilinear one. The practical considerations are to keep the computations simple and accurate. Some examples will serve to illustrate the ideas presented in this section.

8.3.2 Example

The use of the filter to determine unknown constants in a model involves an extension of the dynamic system to include the unknown constants. The following simple example will illustrate the use of the filter to determine the unknown constants. It will also show how the various terms in Equation 8.3.1-5 are derived.

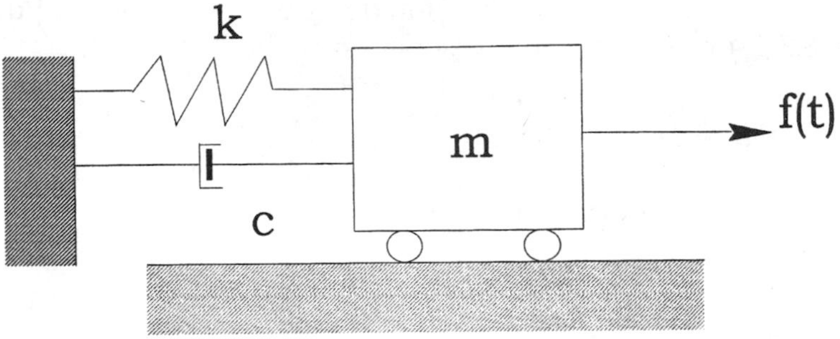

FIGURE 8.3
Simple spring mass system.

Consider the simple spring mass system shown in Figure 8.3 and given by

$$\ddot{x} + 2\zeta\omega\dot{x} + \omega^2 x = 0 \tag{8.3.2-1}$$

where ζ and ω are considered constants to be determined from measurements of either velocity or displacements. For the case considered here, only the displacement measurement will be used. In order to use the filter equations, the constants must be considered variables. Define the following vector

$$\mathbf{x} = \begin{bmatrix} x_1 \\ x_2 \\ x_3 \\ x_4 \end{bmatrix} = \begin{bmatrix} x \\ \dot{x} \\ \zeta \\ \omega \end{bmatrix} \tag{8.3.2-2}$$

The complete dynamic system is given by

$$\begin{aligned} &\dot{x}_1 - x_2 = 0 \\ &\dot{x}_2 + 2x_2x_3x_4 + x_1x_4^2 = 0 \\ &\dot{x}_3 = 0 \\ &\dot{x}_4 = 0 \end{aligned} \tag{8.3.2-3}$$

The matrices **K** and **A** are now filled in with the appropriate terms. The nonlinear vector consists of one term which is given as

$$n_2(x) = 2x_2x_3x_4 + x_1x_4^2 \tag{8.3.2-4}$$

The nonzero entries in the Jacobian matrix **A** are

$$A(2,1) = x_4^2, \qquad A(2,2) = 2x_3x_4$$
$$A(2,3) = 2x_2x_4, \qquad A(2,4) = 2(x_2x_3 + x_1x_4)$$

The identification problem now considers the constants as estimates of state variables. These are to be estimated using only the generated displacement data.

The initial values used were $x_1 = 1.0$, $x_2 = 0.0$, $x_3 = 0.05$, $x_4 = 1000$, the original displacement data and the results of the filter are shown in Figure 8.4. An integration timestep of 2.5E–5 seconds was used for a total time of 0.01 seconds. The filter had no difficulty in following the data. In addition, the velocity has also been reproduced and compared with the original expression (see Figure 8.5). The progress of the constants is shown in Figures 8.6 and 8.7. Both of the constants reach the correct values of 0.1 and 1715 Hz in 0.0015 seconds.

8.4 EXTENDED LEVENBERG-MARQUARDT'S METHOD

In the previous sections, the emphasis was on estimating the entire state vector of the model, and the tacit assumption was that a parameter identification problem would be solved by adjoining the model with equations for the parameters. This idea is only practical when the dimension of the state vector is relatively small. In practice, engineering models can easily reach several thousand state variables. In this section, we would like to outline a method that is applicable to models that are very large yet have a relatively small number of unknown parameters. It is an extension of the Levenberg-Marquardt method, which is normally used for identifying parameters in steady-state models.

Let a general thermal dynamic model be represented as

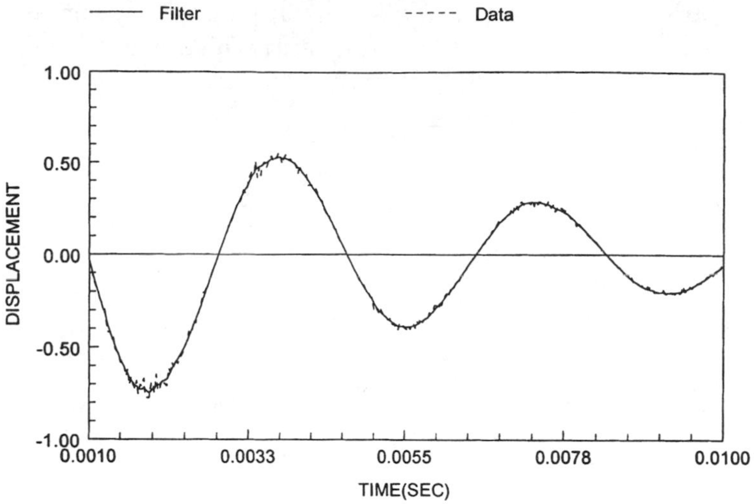

FIGURE 8.4
Comparison of estimated and noisy displacement data.

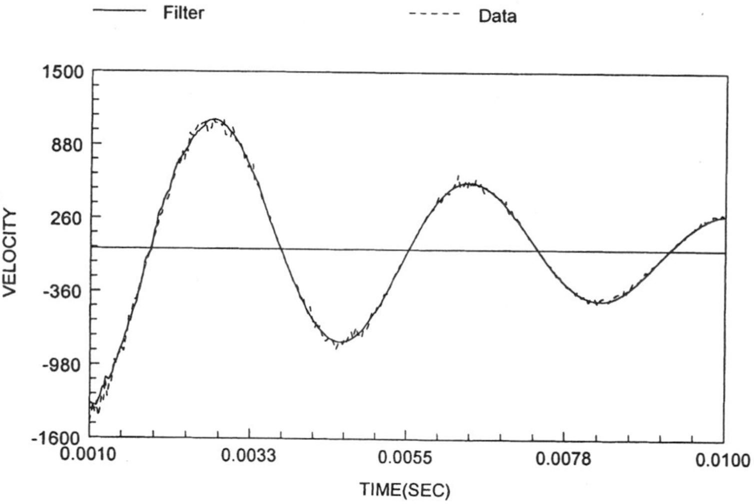

FIGURE 8.5
Comparison of estimated and original velocity.

$$\mathbf{T}_{i+1} = \mathbf{M}_i \mathbf{T}_i + \mathbf{q}_i \tag{8.4-1}$$

where the vector $\mathbf{T}_i$ represents temperatures and may be on the order of several thousand. Let the vector $\mathbf{x}$ represent the

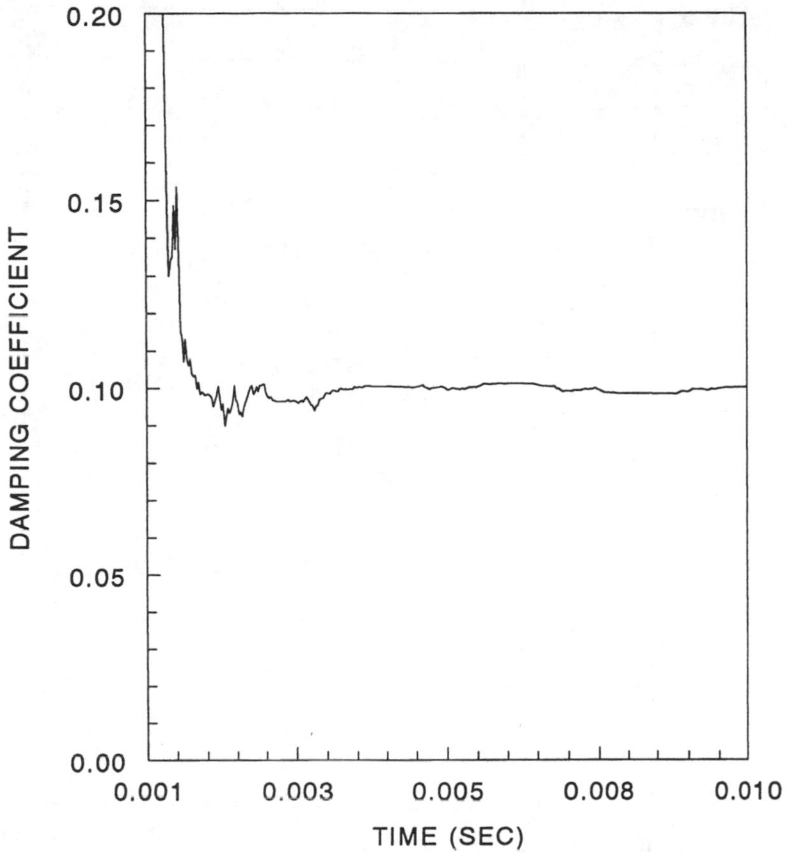

FIGURE 8.6
Convergence of damping coefficient.

unknown parameters. Now the least square error for N data points is

$$E_N(x) = \sum_{i=1}^{N} \left(\mathbf{d}_i - \mathbf{T}_i, \mathbf{d}_i - \mathbf{T}_i\right) + \lambda\left(\mathbf{U}_i\left(\mathbf{x} - \mathbf{x}_0\right), \left(\mathbf{x} - \mathbf{x}_0\right)\right) \quad (8.4\text{-}2)$$

As usual, we have added a regularization term to add stability to the process. The λ constant and the matrix $\mathbf{U}_i$ will be defined below. The data are represented with the vector $\mathbf{d}_i$. Notice that the error depends on $\mathbf{x}$, the unknown parameters.

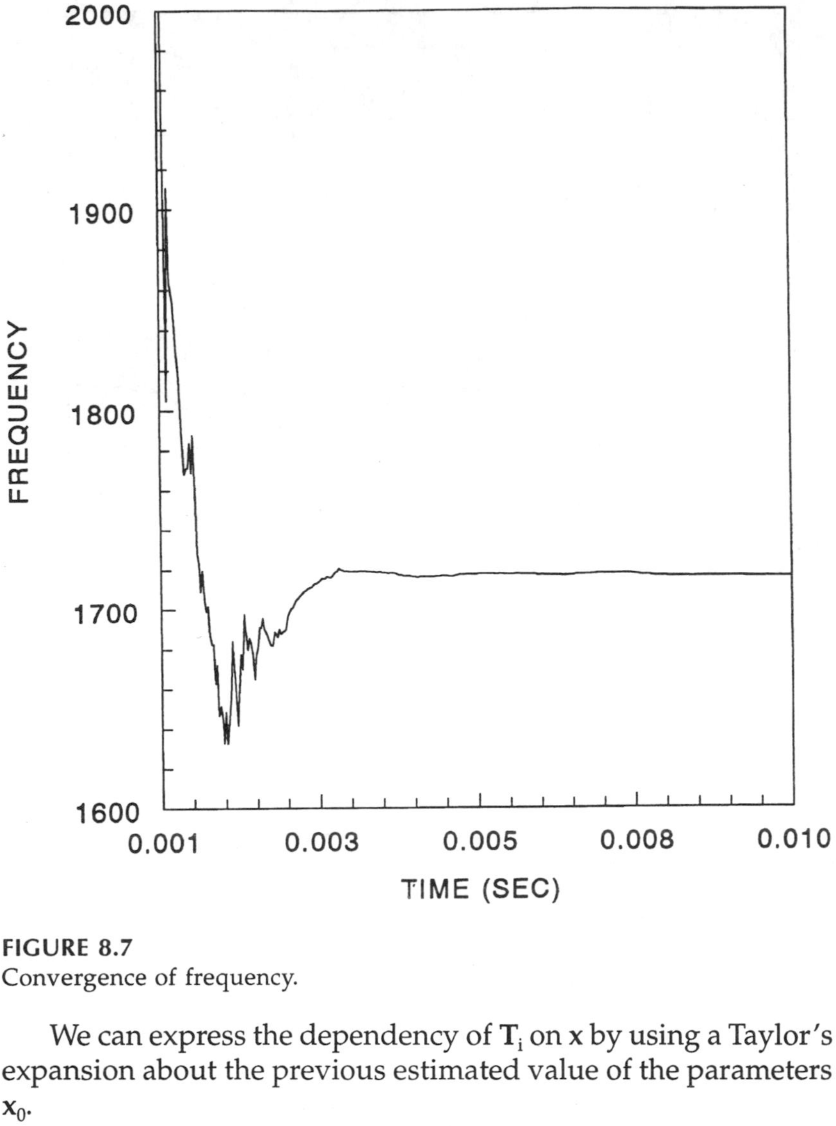

FIGURE 8.7
Convergence of frequency.

We can express the dependency of $\mathbf{T}_i$ on $\mathbf{x}$ by using a Taylor's expansion about the previous estimated value of the parameters $\mathbf{x}_0$.

$$\mathbf{T}_i(\mathbf{x}_0) = \mathbf{T}_i(\mathbf{x}_0) + \mathbf{J}_i(\mathbf{x} - \mathbf{x}_0) \tag{8.4-3}$$

where $\mathbf{J}_i$ represents the Jacobian matrix, which is evaluated at $\mathbf{x}_0$. Substituting this expression into $E_N(\mathbf{x})$ and replacing $\mathbf{x} - \mathbf{x}_0$ with $\mathbf{z}$ (for simplification) gives

$$E_N(x) = \sum_{i=1}^{N} \left(\mathbf{d}_i - \mathbf{T}_i(x_0), \mathbf{d}_i - \mathbf{T}_i(x_0)\right) - 2\left(\mathbf{d}_i - \mathbf{T}_i(x_0), \mathbf{J}_i\mathbf{z}\right) \tag{8.4-4}$$

$$+ \left(\mathbf{J}_i\mathbf{z}, \mathbf{J}_i\mathbf{z}\right) + \lambda\left(\mathbf{U}_i\mathbf{z}, \mathbf{z}\right)$$

The partial derivative of $E_N(x)$ with respect to $\mathbf{z}$ yields a matrix-vector equation

$$\partial E_N(x)/\partial x = \sum_{i=1}^{N} \mathbf{J}_i^T\mathbf{J}_i\mathbf{z} - \mathbf{J}_i^T\left(\mathbf{d}_i - \mathbf{T}_i(x_0)\right) + 2\lambda\mathbf{U}_i\mathbf{z}$$

In the above equations, $\mathbf{J}$ is an $(N_0 \times m)$ matrix where N_0 is the number of temperatures in the model (length of $\mathbf{T}$) and m is the number of parameters to be estimated (length of $\mathbf{x}$). In this section we are interested in the case where N_0 is much larger than m. The Levenberg-Marquardt method sets the matrix $\mathbf{U}_i$ equal to the diagonal of the matrix $\mathbf{J}_i^T\ \mathbf{J}_i$ (Beck and Arnold, 1977).

8.4.1 Sequential Estimation

Sequential estimation is more useful in real-time situations than for the case at hand. However, it can be very useful in truncating the number of data points if the estimates show a strong convergence as the data points are added.

The sequential formulas can be derived by noticing that the error term can be divided into a current part at i = N and the remainder $E_{N-1}(\mathbf{z})$.

$$E_N(\mathbf{z}) = \left(\mathbf{d}_N - \mathbf{T}_N, \mathbf{d}_N - \mathbf{T}_N\right) - 2\left(\mathbf{d}_N - \mathbf{T}_N, \mathbf{J}_N\mathbf{z}\right) \tag{8.4.1-1}$$

$$+ \left(\mathbf{J}_N\mathbf{z}, \mathbf{J}_N\mathbf{z}\right) + \lambda\left(\mathbf{U}_N\mathbf{z}, \mathbf{z}\right) + E_{N-1}(\mathbf{z})$$

It is also recognized that E_N is always a quadratic of the general form

$$E_N(\mathbf{z}) = b_N - 2\left(\mathbf{z}, \mathbf{s}_N\right) + \left(\mathbf{R}_N\mathbf{z}, \mathbf{z}\right) \tag{8.4.1-2}$$

where b_N is a scalar, $\mathbf{s}_N$ is a vector, and $\mathbf{R}_N$ is a matrix. Using these new variables, some recurrence formulas can be derived

$$b_N = b_{N-1} + (\mathbf{d}_N - \mathbf{T}_N, \mathbf{d}_N - \mathbf{T}_N) \quad (8.4.1\text{-}3)$$

$$\mathbf{s}_N = \mathbf{s}_{N-1} + \mathbf{J}_N^T(\mathbf{d}_N - \mathbf{T}_N) \quad (8.4.1\text{-}4)$$

$$\mathbf{R}_N = \mathbf{R}_{N-1} + \mathbf{J}_N^T\mathbf{J}_N + \lambda\mathbf{U}_N \quad (8.4.1\text{-}5)$$

The optimal estimate $\mathbf{z}^*$ can be obtained at any stage N by differentiating the quadratic expression for $E_N(\mathbf{z})$.

$$\partial E_N/\partial \mathbf{z} = -2\mathbf{s}_N + 2\mathbf{R}_N\mathbf{z}^* \quad (8.4.1\text{-}6)$$

and setting it to zero, giving

$$\mathbf{z}_N^* = \mathbf{R}_N^{-1}\mathbf{s}_N \quad (8.4.1\text{-}7)$$

At this point we have all we need for our particular situation. However, this subject has many interesting facets. If one peruses the literature on sequential estimation one will find much discussion of the inverse of $\mathbf{R}_N$ in the above equation. Does it exist? Is it ill-conditioned? Why not avoid computing it? Most of these questions are important to real-time situations or when the number of unknown parameters is much greater than the number of measurements. Our situation is one of the simplest, and experience has shown that the Levenberg-Marquardt technique is a very stable process.

8.4.2 Calculation of the Jacobian

Most of the computational effort is in the calculation of the Jacobian. As an example, assume that a nonlinear thermal model has been simulated with a fully implicit integration scheme. The complete equation is

$$\mathbf{C}(\mathbf{T}_{i+1} - \mathbf{T}_i)/h + \mathbf{K}_i\mathbf{T}_{i+1} + \mathbf{K}_r\mathbf{T}_{i+1}^4 = \mathbf{q}_i \quad (8.4.2\text{-}1)$$

where $\mathbf{C}_i$ represents the capacitance matrix, $\mathbf{K}_i$, the conductance matrix, $\mathbf{K}_r$, the radiation matrix, and $\mathbf{q}_i$, the applied heat fluxes. The Jacobian matrix is the partial derivative of this equation with respect to the unknown parameters $\mathbf{x}$. Letting

$$\mathbf{J}_i = [\partial \mathbf{T}_i / \partial \mathbf{x}] \tag{8.4.2-2}$$

$$\mathbf{J}_{i+1} = [\partial \mathbf{T}_{i+1} / \partial \mathbf{x}] \tag{8.4.2-3}$$

gives, *symbolically,* for the parameters affecting the linear matrix, the equation

$$\mathbf{C}\mathbf{J}_{i+1}/h + \mathbf{K}_i\mathbf{J}_{i+1} + 4\mathbf{K}_r\mathbf{T}_{i+1}^3\mathbf{J}_{i+1} = \mathbf{C}\mathbf{J}_i/h - [\partial \mathbf{K}_i/\partial \mathbf{x}]\mathbf{T}_{i+1} \tag{8.4.2-4}$$

The unknowns $\mathbf{J}_{i+1}$ have been grouped on the left side of the equation and the known quantities on the right side. If the unknown parameters affect the capacitance matrix, the above equations would become

$$\begin{gathered} \mathbf{C}_i\mathbf{J}_{i+1}/h + \mathbf{K}_i\mathbf{J}_{i+1} + 4\mathbf{K}_r\mathbf{T}_{i+1}^3\mathbf{J}_{i+1} = \\ \mathbf{C}_i\mathbf{J}_i/h - [\partial \mathbf{C}_i/\partial \mathbf{x}][\mathbf{T}_{i+1} - \mathbf{T}_i]/h \end{gathered} \tag{8.4.2-5}$$

An important characteristic of the above equations is that they must be simultaneously integrated with the thermal model since they depend on the temperatures $\mathbf{T}_{i+1}$ and $\mathbf{T}_i$.

There are several strategies that can be used to control the updating of the unknown parameters. The simplest (Press et. al., 1990) is to first choose λ to be a small value (0.001) and then increase it or decrease it by a factor of 10 depending on whether or not the updated parameters increased or decreased the overall error. This updating is performed only after all the data points have been included in the estimate. In some special very non-linear cases, the sequential estimator can be fully extended to include updating the model parameters with every time step. The user is encouraged to try different strategies for his particular problem.

BIBLIOGRAPHY

This book was intended to be as self-contained as possible. References are cited so that the reader may find additional information on the subjects presented. No attempt was made to supply a comprehensive bibliography. There has been a tremendous amount of research in some of the subjects discussed and the authors apologize in advance for omissions.

Aho, A., Hopcroft, J.E., and Ullman, J.D., *The Design and Analysis of Computer Algorithms,* Addison-Wesley, Reading, Massachusetts, 1974.

Alifanov, O.M., Derivation of formulas for the gradient of the error in the iterative solution of inverse problems of heat conduction. I. Determination of the gradient in terms of the Green's function, *J. Eng. Phys.*, 52, 352, 1987.

Alifanov, O.M., Methods of solving ill-posed inverse problems, *J. Eng. Phys.*, 45, 1237, 1983.

Alifanov, O.M. and Egorov, Yu.V., Algorithms and results of solving the inverse heat-conduction boundary problem in a two-dimensional formulation, *J. Eng. Phys.*, 48, 489, 1985.

Alifanov, O.M. and Kerov, N.V., Determination of external thermal load parameters by solving the two-dimensional inverse heat-conduction problem, *J. Eng. Phys.*, 41, 1049, 1982.

Angel, E., Dynamic programming and linear partial differential equations, *J. Math. Anal. Appl.*, 23, 628, 1968.

Anger, G., *Inverse Problems in Differential Equations,* Plenum Press, New York, 1990.

Arledge, R.G. and Haji-Sheikh, A., An interactive approach to the solution of inverse heat conduction problems, ASME Paper No. 77-WA/TM-2, 1977.

Astrom, K.J. and Eykhoff, P., System identification-A survey, *Automatica,* 7, 123, 1971.

Baranenko, V.A. and Pochtman, Iu.M., Dynamic programming and nonlinear satics problems of thin rods, *Dokl. Akad. Nauk. SSSR,* 182, 5, 1968.

Baranenko, V.A. and Pochtman, Iu.M., Investigation of the deformation of elastic membranes constrained by restrictions (on displacements) by the method of dynamic programming, *PMM,* 33, 5, 933, 1969.

Bass, B.R., Application of the finite element method to the nonlinear inverse heat conduction problem using Beck's second method, *J. Eng. Ind.*, 102, 168, May 1980.

Bass, B.R. and Ott, L.J., Application of the finite element method to the two-dimensional nonlinear inverse heat conduction problem, in *Numerical Methods for Non-Linear Problems,* C. Taylor, E. Hinton, and D.R.J. Owen, Eds., 1980, 649.

Bathe, K.J. and Wilson, E.L., Stability and accuracy analysis of direct integration methods, *Int. J. Earthq. Engng. Struct. Dyn.,* 1, 283, 1973.

Bathe, K.J. and Wilson, E.L., *Numerical Methods in Finite Element Analysis,* Prentice-Hall, Englewood Cliffs, New Jersey, 1976.

Beck, J.V., Nonlinear estimation applied to the nonlinear inverse heat conduction problem, *Int. J. Heat Mass Transfer,* 13, 703, 1970.

Beck, J.V., Criteria for comparison of methods of the inverse heat conduction problem, *Nucl. Eng. Des.,* 53, 11, 1979.

Beck, J.V., and Arnold, K.J., *Parameter Estimation In Engineering and Science,* John Wiley & Sons, New York, 1977.

Beck, J.V., and Murio, D.A., Combined function specification-regularization procedure for solution of inverse heat conduction problem, *AIAA Journal,* 24, 1, 180, January 1986.

Beck, J.V., Blackwell, B., and St. Clair, Jr., C.R., *Inverse Heat Conduction Ill-Posed Problems,* Wiley Interscience, New York, 1985.

Beck, J.V., Litkouhi, B., and St. Clair, Jr., C.R., Efficient sequential solution of the nonlinear inverse heat conduction problem, *Numer. Heat Trans.,* 5, 275, 1982.

Bellman, R., A variational problem with constraints in dynamic programming, *J. Soc. Indust. Appl. Math.,* 4, 1, 48, March 1956.

Bellman, R., *Dynamic Programming,* Princeton University Press, New Jersey, 1957.

Bellman, R., Dynamic programming, invariant imbedding, and two-point boundary value problems, *Boundary Problems in Differential Equations,* University of Wisconsin Press, Madison, pp. 257-272, 1960a.

Bellman, R., *Introduction to Matrix Analysis,* McGraw-Hill, New York, 1960b.

Bellman, R., *Introduction to the Mathematical Theory of Control Processes,* Vols. 1 and 2, Academic Press, New York, 1967.

Bellman, R., *Methods of Nonlinear Analysis,* Vols. I and II, Academic Press, New York and London, 1973.

Bellman, R., *Mathematical Methods in Medicine,* World Scientific Publishing, Singapore, 1983.

Bellman, R. and Dreyfus, S., *Applied Dynamic Programming,* Princeton University Press, New Jersey, 1962.

Bellman, R. and Kalaba, R., *Dynamic Programming and Modern Control Theory,* Academic Press, New York, 1965.

Bellman, R. and Roth, R., *Quasilinearization and the Identification Problem,* World Scientific Publishing, Singapore, 1983.

Bellman, R., Kagiwada, H., Kalaba, R., and Sridhar, R., Invariant imbedding and nonlinear filtering theory, *J. Astronaut. Sci.,* XIII, 3, 110, 1966.

Belytschko, T. and Hughes, T.J.R., Eds., *Computational Methods for Transient Analysis,* Amsterdam, North-Holland, 1983.

Berman, A., Determining structural parameters from dynamic testing, *Shock Vibr. Dig.,* 7, 10, 1975.

Bierman, G.J., *Factorization Methods for Discrete Sequential Estimation*, Academic Press, New York, 1977.

Bryson, Jr., A.E. and Ho, Y-C., *Applied Optimal Control Optimization, Estimation, and Control*, Hemisphere Publishing Corporation, Washington DC, 1975.

Budgell, W.P., Stochastic filtering of linear shallow water wave processes, *SIAM J. Sci. Comput.*, 8, 2, 1987.

Bulychev, E.V. and Glasko, V.B., Uniqueness in certain inverse problems of the theory of heat conduction, *J. Engr. Phys.*, 45, 940, 1983.

Burggraf, O.R., An exact solution of the inverse problem in heat conduction theory and applications, *ASME J. Heat Trans.*, 373, August 1964.

Busby, H.R. and Trujillo, D.M., Numerical experiments with a new differentiation filter, *ASME J.Biomech. Eng.*, 107, 293, November 1985a.

Busby, H.R. and Trujillo, D.M., Numerical solution to a two-dimensional inverse heat conduction problem, *Int. J. Numer. Meth. Eng.*, 21, 349, 1985b.

Busby, H.R. and Trujillo, D.M., Solution of an inverse dynamics problem using an eigenvalue reduction technique, *Computers and Structures*, 25, 1, 109, 1987.

Busby, H.R., Dohrmann, C.R., and Trujillo, D.M., Frequency analysis of an optimal smoothing and differentiating filter, *Mech. Sys. Signal Proc.*, 3, 4, 361, 1989.

Cannon, J.R. and Hornung, U., (Eds.), *Inverse Problems*, Birkhauser Verlag, Basel, 1986.

Casti, J., The linear-quadratic control problem: some recent results and outstanding problems, *SIAM Rev.*, 22, 4, 459, October 1980.

Chan, S.P., Cox, H.L., and Benfield, W.A., Transient analysis of forced vibrations of complex structural-mechanical systems, *J. Roy. Aeron. Soc.*, 66, 457, 1962.

Chiu, L.H. and Schumaker, L.L., Complete spline smoothing, *Numer. Math.*, 49, 1, 1986.

Clough, R.W. and Penzien, J., *Dynamics of Structures*, McGraw-Hill, New York, 1975.

Cooper, L. and Cooper, M.W., *Introduction to Dynamic Programming*, Pergamon Press, Oxford, 1981.

Craven, P. and Wahba, G., Smoothing noisy data with spline functions, *Numer. Math.*, 31, 377, 1979.

D'Cruz, J., Crisp, J.D.C., and Ryall, T.G., On the identification of a harmonic force on a viscoelastic plate from response data, *ASME J. Appl. Mech.*, 59, 722, December 1992.

Detchmendy, D.M. and Sridhar, R., Sequential estimation of states and parameters in noisy nonlinear dynamical systems, *ASME J. Basic Eng.*, 8, 2, 362, June 1966.

Deverall, L.I. and Channapragada, R.S., A new integral equation for heat flux in inverse heat conduction, *ASME J. Heat Trans.*, 327, August 1966.

Distefano, N., Dynamic programming and invariant imbedding in structural mechanics, in *Invariant Imbedding*, Edited by R. Bellman and E.D. Denman, Springer-Verlag, Berlin, 1971, 118.

Distefano, N., *Nonlinear Processes in Engineering*, Academic Press, New York, 1974a.

Distefano, N. and Rath, A., *Modeling and Identification in Nonlinear Structural Dynamics-I. One Degree of Freedom Models,* Earthquake Engineering Research Center, University of California, Berkeley, Report No. EERC 74-15, December 1974b.

Dohrmann, C.R. and Busby, H.R., Algorithms for smoothing noisy data with spline functions and smoothing parameter selection, VI International Congress on Experimental Mechanics, Portland Oregon, June 5-10, 1988, 843.

Dohrmann, C.R. and Busby, H.R., Spline function smoothing and differentiation of noisy data on a rectangular grid, 1990 SEM Spring Conference on Experimental Mechanics, Albuquerque, New Mexico, June 4-6, 1990, 76.

Dohrmann, C.R., Trujillo, D.M., and Busby, H.R., Smoothing noisy data using dynamic programming and generalized cross-validation, *ASME J. Biomech. Eng.*, 37, February 1988.

Dreyfus, S., Dynamic programming and the calculus of variations, *J. Math. Anal. App.*, 1, 228, 1960.

Dreyfus, S. and Kan, Y.C., A general dynamic programming solution of discrete-time linear optimal control problems, *IEEE Transactions on Automatic Control,* June 1973, 286.

Elbert, T.F., *Estimation and Control of Systems,* Van Nostrand Reinhold, New York, 1984.

Eykhoff, P., *System Identification, Parameter and State Estimation,* John Wiley & Sons, New York, 1974.

Fine, N.J., The Jeep problem, *Amer. Math. Monthly,* LIV, January, 1947.

Frank, I., An application of least squares method to the solution of the inverse problem of heat conduction, *ASME* J. Heat Trans., 378, November 1963.

Fritzen, C-P, Identification of mass, damping, and stiffness matrices of mechanical systems, *J. Vibr., Acoust., Stress, and Reliability Des.*, 108, 9, January 1986.

Fuchs, H., Kedem, Z.M., and Uselton, S.P., Optimal surface reconstruction using planar contours, *Comm. ACM,* 20, 10, 693, 1971.

Galiullin, A.S., *Inverse Problems of Dynamics,* Mir Publishers Moscow, 1984.

Gersch, W., Parameter identification:stochastic process techniques, *Shock Vibr. Dig.*, 7, 71, 1975.

GhoshRay, D.N., *Methods of Inverse Problems in Physics,* CRC Press, Boca Raton, 1991.

Gladwell, G.M.L., Inverse problems in vibration, *Appl. Mech. Rev.*, 39, 7, 1013, July 1986.

Golub, G.H., Heath, M., and Wahba, G., Generalized cross-validation as a method for choosing a good ridge parameter, *Technometrics,* 21, 215, May 1979.

Haario (Ed.), *Theory and Applications of Inverse Problems,* Longman Scientific & Technical, Longman Group UK, 1988.

Hansen, P.C., Analysis of discrete ill-posed problems by means of the l-curve, *SIAM Rev.*, 34, 4, 561, December 1992.

Hart, G.C. and Yao, J.T.P., System identification in structural dynamics, *ASCE J. Eng. Mech. Div.*, 103, EM6, 1089, December 1977.

Hensel, E., *Inverse Theory and Applications for Engineers,* Prentice Hall, Englewood Cliffs, New Jersey, 1991.

Hensel, E., Inverse problems for multi-dimensional parabolic partial differential equations, *Appl. Mech. Rev.*, 41, 6, 263, June 1988.

Hensel, E. and Hills, R.G., An initial value approach to the inverse heat conduction problem, *ASME J. Heat Trans.*, 108, 248, May 1986.

Hills, R.G., Mulholland, G.P., and Matthews, L.K., The application of the backus-gilbert method to the inverse heat conduction problem in composite media, *ASME* Paper No. 82-HT-26, 1982.

Hohn, F.E., *Elementary Matrix Algebra*, Macmillan, New York, 1957.

Hollandsworth, P.E. and Busby, H.R., Impact force identification using the general inverse technique, *Int. J. Impact Eng.*, 8, 4, 315, 1989.

Hsu, Y.F., Rubinsky, B., and Mahin, K., An inverse finite element method for the analysis of stationary arc welding processes, *ASME J. Heat Trans.*, 108, 734, November 1986.

Huan, S.-L., McInnis, B.C., and Denman, E.D., Identification of structural systems using naturally induced vibration data in the presence of measurement noise, *Comp. Meth. Appl. Mech. Eng.*, 41, 123, 1983.

Hughes, T.R.J., *The Finite Element Method—Linear Static and Dynamic Finite Element Analysis*, Prentice Hall, Englewood Cliffs, New Jersey, 1987.

Hutchinson, M.F. and de Hoog, F.R., Smoothing noisy data with spline functions, *Numer. Math.*, 47, 99, 1985.

Ibanez, P., Identification of dynamic parameters of linear and non-linear structural models from experimental data, *Nucl. Eng. Des.*, 25, 30, 1973.

Ibanez, P., methods for the identification of dynamic parameters of mathematical structural models from experimental data, *Nucl. Eng. Des.*, 27, 209, 1974.

Imber, M., Nonlinear heat transfer in planer solids: direct and inverse applications, *AIAA Journal*, 17, 2, 204, February 1979.

INTEMP, A Computer Program for Nonlinear Inverse Heat Conduction, TRUCOMP, Fountain Valley, CA, 1988.

Isaacson, E. and Keller, H.B., *Analysis of Numerical Methods*, John Wiley & Sons, New York, 1966.

Iskenderov, A.D., Dzhafarov, Dzh.F., and Gumbatov, O.A., Explicit solutions of multidimensional inverse unsteady heat-conduction problems, *J. Eng. Phys.*, 41, 778, 1982.

Jacobson, D.H. and Mayne, D.Q., *Differential Dynamic Programming*, American Elsevier, New York, 1970.

Jennings, A., *Matrix Computations for Engineers and Scientists*, John Wiley & Sons, New York, 1978.

Kagiwada, H., Kalaba, R.E., Schumitzky, A., and Sridhar, R., Invariant Imbedding and sequential interpolating filters for nonlinear processes, *ASME J. Basic Eng.*, 91, 2, 195, June 1969.

Kalaba, R., Dynamic programming and the variational principles of classical and statistical mechanics, in *Developments in Mechanics*, Vol. 1, Proc. of the Seventh Midwestern Mechanics Conference, Edited by J.E. Lay and L.E. Malvern, Plenum Press, New York, 1961.

Kalaba, R. and Spingran, K., *Control, Identification, and Input Optimization*, Plenum Press, New York, 1982.

Kalaba, R. and Tesfatsion, L., An exact sequential solution procedure for a class of discrete-time nonlinear estimation problems, *IEEE Transactions on Automatic Control*, AC-26, 5, 1144, October 1981.

Kalaba, R. and Tesfatsion, L., Exact sequential filtering, smoothing and prediction for nonlinear systems, *Nonlinear Anal., Theory, Meth. Appl.*, 12, 6, 599, 1988.

Kerov, N.V., Solution of the two-dimensional inverse heat-conduction problem in a cylindrical coordinate system, *J. Engr. Phys.*, 45, 1245, 1983.
Klein, S., and Trujillo, D.M., An unconditionally stable finite element analysis for nonlinear structures, *Computers and Structures*, 16, 1-4, 187, 1983.
Knight, J.H. and Phillip, J.R., Exact solutions in nonlinear diffusion, *J. Eng. Math.*, 8, 219, 1974.
Knuth, D.E., Optimal binary search trees, *Acta Informatica*, 1:1, 14, 1971.
Kosarev, A.A., Milovskaya, L.S., and Cherpakov, P.V., Modeling reverse problems of heat conduction with moving phase transition boundaries, *J. Eng. Phys.*, 44, 692, 1983.
Kozdoba, L.A., Basis of the terminology and algorithm for the solution of inverse heat-transfer problems, *J. Eng. Phys.*, 45, 1322, 1983.
Krutz, G.W., Schoenhals R.J., and Hore, P.S., Application of the finite-element method to the inverse heat conduction problem, *Numerical Heat Transfer*, 1, 489, 1978.
Krzysztof, G., Cialkowski, M.J., and Kaminski, H. An inverse temperature field problem of the theory of thermal stresses, *Nucl. Eng. Des.*, Vol. 64, 1981
Kubo, S., Inverse problems related to the mechanics and fracture of solids and structures, *JSME Int. J.*, Series I, 31, 2, 157, 1988.
Lanczos, C., *Applied Analysis*, Prentice Hall, Englewood Cliffs, New Jersey, 1964.
Lapidus, L. and Luus, R., *Optimal Control of Engineering Processes*, Blaisdell Publishing Company, Waltham Massachusetts, 1967.
Larson, R.E. and J.L. Casti, J.L., *Principles of Dynamic Programming*, Part 1 and 2, Dekker Publishing Co., 1982.
Lawson, C.L. and Hanson, R.J., *Solving Least Squares Problems*, Prentice-Hall, Englewood Cliffs, New Jersey, 1974.
Lazuchenkov, N.M. and Shmukin, A.A., Inverse boundary-value problem of heat conduction for a two-dimensional domain, *J. Eng. Phys.*, 40, 223, 1981.
Lim, T.W. and Pilkey, W.D., A solution to the inverse dynamics problem for lightly damped flexible structures using a modal approach, *Computers and Structures*, 43, 53, 1992.
Lin, P.P. and Datseris, P., Inverse problem solutions via plate theory with applications to position and force sensing, *J. Appl. Mech.*, Vol. 55, 1988.
Maiorov, A.I. and Rudometkin, L.A., Determining the mean-square error and discretization step of the initial data of an inverse problem in a single realization, *J. Eng. Phys.*, 51, 876, 1987.
Maniatty, A., Zabaras, N., and Stelson, K., Finite element analysis of some inverse elasticity problems, *ASCE J. Eng. Mech.*, 115, 6, 1303, June 1989.
Manoach, E., Karagiozova, D., and Hadjikov, L., An inverse problem for an initially heated circular plate under a pulse loading, *ZAMM*, Vol., 1991.
Markovsky, A., Development and application of ill-posed problems in the USSR, *Appl. Mech. Rev.*, 41, 6, 247.
Martinez, Y. and Dinh, P.N., A generalization of Tikhonov's regularizations of zero and first order, *Comp. & Maths. with Appls.*, 12B, 5/6, 1203, 1986.
Matsevityi, Yu.M., One approach to solving reverse problems of heat conduction, *J. Eng. Phys.*, 41, 1384, 1982.
Matsevityi, Yu.M., Identification of thermophysical characteristics by solving inverse heat conductivity problems, *J. Eng. Phys.*, 45, 1165, 1983.
Mayne, D.Q., Parameter estimation, *Automatica*, 3, 245, 1966a.

Mayne, D.Q., A solution of the smoothing problem for linear dynamic systems, *Automatica*, 4, 73, 1966b.

Mendel, J.M., *Discrete Techniques of Parameter Estimation—The Equation Error Formulation*, Marcel Dekker, New York, 1973.

Meric, R.A., Finite element and conjugate gradient methods for a nonlinear optimal heat transfer control problem, *Int. J. Numer. Meth. Eng.*, 14, 1851, 1979.

Metha, R.C., Numerical solution of nonlinear inverse heat conduction problem with a radiation boundary condition, *Int J. Numer. Meth. Eng.*, 20, 1057, 1984.

Michaels, J.E. and Pao, Yih-Hsing, The inverse source problem for an oblique force on an elastic plate, *J. Acoust. Soc. Am.*, 77, 6, 1985.

Moler, C. and Loan, C.V., Nineteen dubious ways to compute the exponential of a matrix, *SIAM Rev.*, 20, 801, 1978.

Mook, D.J., Estimation and identification of nonlinear dynamic systems, *AIAA Journal*, 27, 7, 968, July 1989.

Morf, M., Sidhu, G., and Kailath, T., Some New algorithms for recursive estimation in constant, linear, discrete-time systems, *IEEE Trans. Automatic Control*, AC-19, 4, 315, August 1974.

Mulholland, G.P., Gupta, B.P., and San Martin, R.L., Inverse problem of heat conduction in composite media, ASME Paper No. 75-WA/HT-83, 1975.

Murray, D.M., The adjustment and integration of accelerometer data using steady-state feedback gains, *Optimal Control Applications & Methods*, 6, 201, 1985.

Murray, D.M., The relationship between terminal state constraints and penalties for the discrete-time LQP problem associated with the adjustment of accelerometer data, *J. Comput. Appl. Math.*, 18, 83, 1987.

Norton, J.P., *An Introduction to Identification*, Academic Press, London, 1986.

Novikov, I.A., Solution of the linear one-dimensional inverse heat-conduction problem on the basis of a hyperbolic equation, *J. Eng. Phys.*, 40, 668, 1981.

Oda, J. and Shinada, T., An inverse technique to obtain contact stress distribution, *Jpn. Soc. Mech. Eng.*, (in Japanese), 53, 492, 1614, A, 1987.

Osman, A.M. and Beck, J.V., Nonlinear inverse problem for the estimation of time-and-space-dependent heat-transfer coefficients, *AIAA J. Thermophysics*, 3, 2, 146, April 1989.

Palmer, A.C., Optimal structure design by dynamic programming, *ASCE J. Struct. Div.*, 94, ST8, 1887, August 1968.

Palmer, A.C., Limit analysis of cylindrical shells by dynamic programming, *Int. J. Solids Structures*, 5, 289, 1969.

Papadopoulos, A. and Trujillo, D.M., Natural frequency of a Timoshenko beam on a flexible base, *ASCE J. Eng. Mech. Div.*, 307, April 1980.

Pokhoriler, V.L., Solution of two-dimensional heat-conduction problems by the Z transform method, *J. Eng. Phys.*, 44, 564, 1983.

Press, W.H., Flannery, B.P., Teukolsky, S.A., and Vetterling, W.T., *Numerical Recipes*, Cambridge University Press, 1990.

Raynaud, M. and Beck, J.V., Methodology for comparison of inverse heat conduction methods, *ASME J. Heat Trans.*, 110, 30, February 1988.

Reddy, J.N., *An Introduction to the Finite Element Method*, McGraw-Hill, New York, 1984.

Roth, R.S., Techniques in the identification of deterministic systems, *IEEE Transactions on Automatic Control*, AC-26, 5, 1169, October 1981.

Sabater, P.C. (Ed.), *Inverse Methods in Action*, Proceedings of the Multicentennials Meeting on Inverse Problems, Montpellier, November 27th-December 1st, 1989, Springer-Verlag, Berlin, 1990.

Sage, A.P. and Melsa, J.L., *System Identification*, Academic Press, New York, 1971.

Scott, E.P. and Beck, J.V., Analysis of order of sequential regularization solution of inverse heat conduction problem, *ASME Paper* 85-WA/HT-45, 1985.

Shih, T.M. and Skladany, J.T., An eigenvalue method for solving transient heat conduction problems, *Numerical Heat Transfer*, 6, 409, 1983.

Simonian, S.S., Inverse problems in structural dynamics-I. Theory, *Int. J. Numer. Meth. Eng.*, 17, 357, 1981a.

Simonian, S.S., Inverse problems in structural dynamics-II. Applications, *Int. J. Numer. Meth. Eng.*, 17, 367, 1981b.

Sparrow, E.M., Haji-Sheikh, A., and Lundgren, T.S., The inverse problem in transient heat conduction, *J. Appl. Mech.*, Vol. 86, 1975.

Stolz, G., Jr., Numerical solution to an inverse problem of heat conduction for simple shapes, *J. Heat Transfer*, Trans. ASME, Vol. 82, 1960.

Tandy, D.F., Trujillo, D.M., and Busby, H.R., Solution of inverse heat conduction problems using an eigenvalue reduction technique, *Numer. Heat Trans.*, 10, 597, 1986.

Thomson, W.T., Parameter uncertainty in dynamic systems, *Shock Vibr. Dig.*, 7, 3, 1975.

Thornhill, R.J. and Smith, C.C., Impact force prediction using measured frequency response functions, *J. Dyn. Sys. Meas. Contr.*, 105, 227, December 1983.

Tikhonov, A.N. and Arsenin, V.Y., *Solution of Ill-Posed Problems*, Winston and Sons, Washington, D.C., 1977.

Trujillo, D.M., The direct numerical integration of linear matrix differential equations using Pade approximations, *Int. J. Numer. Meth. Eng.*, 9, 259, 1975.

Trujillo, D.M., An unconditionally stable explicit algorithm for finite element heat conduction analysis, *Nucl. Eng. Des.*, 41, 175, 1977a.

Trujillo, D.M., An unconditionally stable explicit algorithm for structural dynamics, *Int. J. Numer. Meth. Eng.*, 11, 1579, 1977b.

Trujillo, D.M., Application of dynamic programming to the general inverse problem, *Int. J. Numer. Meth. Eng.*, 12, 613, 1978.

Trujillo, D.M., Stability analysis of the force correction method for nonlinear structural dynamics, *ASME J. Appl. Mech.*, 49, 203, March 1982.

Trujillo, D.M., and Busby, H.R., Finite element nonlinear heat transfer analysis using a stable explicit method, *Nucl. Eng. Des.*, 44, 227, 1977.

Trujillo, D.M., and Busby, H.R., Investigation of highly accurate integration formulas for transient heat conduction analysis using conjugate gradient technique, *Int. J. Numer. Meth. Eng.*, 18, 99, 1982.

Trujillo, D.M., and Busby, H.R., Investigation of a technique for the differentiation of empirical data, *ASME J. Dyn. Sys. Meas. Contr.*, 105, 200, 1983.

Trujillo, D.M., and Busby, H.R., Optimal regularization of the inverse heat-conduction problem, *AIAA J. Thermophys. Heat Trans.*, 3, 4, 423, October 1989.

Trujillo, D.M., and Busby, H.R., A mathematical method for the measurement of bone motion with skin-mounted accelerometers, *ASME J. Biomech. Eng.*, 112, 229, March 1990.

Trujillo, D.M. and Busby, H.R., Optimal regularization of the inverse heat conduction problem using the L-curve, *Int. J. Num. Meth. Heat Fluid Flow,* 4, 447, 1994.

Trujillo, D.M., and Carter, A.L., A new approach to the integration of accelerometer data, *Earthq. Eng. Struct. Dyn.*, 10, 529, 1982.

Trujillo, D.M., and Wallis, R., Determination of heat transfer from components during quenching, *Ind. Heating,* 22, July 1989.

Tryanin, A.P., Identification of heat-transfer coefficients in a porous body from the solution of an inverse problem, *J. Eng. Phys.*, 45, 1301, 1983.

Vabishchevich, P.N. and Pulatov, P.A., Numerical solution of an inverse heat-conduction boundary problem, *J. Eng. Phys.*, 51, 1097, 1987.

Varga, R.S., *Matrix Iterative Analysis,* Prentice-Hall, London, 1962.

Voskoboinikov, Yu.E. and Preobrazhenskii, N.G., Descriptive solution of the inverse heat-conduction problem in the B-spline basis, *J. Eng. Phys.*, 45, 1254, 1983.

Wahba, G., *Spline Models for Observational Data,* Society for Industrial and Applied Mathematics, Philadelphia, 1990.

Weiland, E. and Babary, J.P., Comparative study for a new solution to the inverse heat conduction problem, *Comm. Appl. Numer. Meth.*, 4, 687, 1988.

Wells, W.R., Stochastic parameter estimation for dynamic systems, *Shock Vibr. Dig.*, 7, 86, 1975.

Woltring, H.J., On optimal smoothing and derivative estimation from noisy displacement data in biomechanics, *Hum. Move. Sci.*, 4, 3, 229, September 1985.

Woltring, H.J., A FORTRAN package for generalized, cross-validatory spline smoothing and differentiation, *Adv. Eng. Software,* 8, 2, 104, 1986.

Yoshimura, T. and Ikuta, K., Inverse heat-conduction problem by finite-element formulation, *Int. J. Sys. Sci.*, 16, 11, 1365, 1985.

Yun, C-B. and Shinozuka, M., Identification of nonlinear structural dynamic systems, *J.Struct. Mech.*, 8, 2, 187, 1980.

APPENDIX A

DYNAVAL* USER'S MANUAL

A.1 INTRODUCTION

DYNAVAL solves the general inverse problem for any model that can be represented by a set of linear differential equations. The inverse problem is one in which some of the state variables have been measured and the forcing functions are unknown. DYNAVAL combines the measurements and the model to optimally estimate the forcing functions. This difficult problem is solved using the methods of dynamic programming and generalized cross validation (GCV). The input to the program consists of model definition, relationship of the state variable to the measurements, and the measurements themselves. The input to the program is in the form of a free field format.

DYNAPLOT provides graphics post-processing that will plot the results of DYNAVAL. Time history plots of the state variables, the forcing functions, and the measurements are available.

* A Computer Program for the Solution of the General Inverse Problem Using Dynamic Programming and Generalized Cross-Validation
August, 1995

TRUCOMP, 18571 Las Flores, Fountain Valley, CA, 92708

Also, the measurements can be compared to the model results. DYNAPLOT is self explanatory and prompts the user for the necessary information.

A.2 ANALYSIS FORMULATION

DYNAVAL uses the following vector-matrix model:

$$\mathbf{C}\dot{\mathbf{x}} = \mathbf{K}\mathbf{x} + \mathbf{P}\mathbf{g} \tag{A-1}$$

where

$\mathbf{x}$ = state variables ($N \times 1$)
$\mathbf{C}$ = diagonal matrix ($N \times N$)
$\mathbf{K}$ = system matrix ($N \times N$)
$\mathbf{g}$ = unknown forcing terms ($N_G \times 1$)
$\mathbf{P}$ = influence matrix ($N \times N_G$)

The matrices **C**, **K**, and **P** are constant. Equation (A-1) is converted to discrete form using the exponential matrix (see Section 2.3)

$$\mathbf{x}_{i+1} = \mathbf{M}\mathbf{x}_i + \mathbf{P}^* \mathbf{g}_i \tag{A-2}$$

The measurements at each sample of time are

$\mathbf{d}_i$ = measurement vector at time ih ($N_Z \times 1$)
h = time step

These are related to the state variables with a general relationship of the form

$$\mathbf{z}_i = \mathbf{U}\mathbf{x}_i \tag{A-3}$$

where

$\mathbf{U}$ = measurement influence matrix ($N_Z \times N$)
$\mathbf{z}_i$ = vector to be compared with $\mathbf{d}_i$ ($N_Z \times 1$)

DYNAVAL offers an option that will allow the user to input the matrices **M** and **P** directly.

The general inverse problem is to find the unknown forcing terms $\mathbf{g}_i$ that closely match the measured data. This is accomplished by minimizing the least squares error function E

$$E = \sum_{i=1}^{n} \left(\mathbf{z}_i - \mathbf{d}_i, \mathbf{A}(\mathbf{z}_i - \mathbf{d}_i)\right) + \left(\mathbf{g}_i, \mathbf{B}\mathbf{g}_i\right) \qquad \text{(A-4)}$$

where

A = measurement weighting matrix ($N_Z \times N_Z$)
B = forcing term weighting matrix ($N_G \times N_G$)

A is usually the identity matrix (**I**) and **B** is a relative matrix which is multiplied by a smoothing parameter b. The reason for this restriction is that the parameter b will be selected in an optimal manner using Generalized Cross-Validation.

A.3 INPUT DATA INSTRUCTIONS

Card A	Title
Card B	Control Card
Card C	Time Data
Card D	Matrix **C** (Eq. A-1)
Card E	Matrix **K** (Eq. A-1)
Card F	Matrix **P** (Eq. A-1)
Card G	Matrix **U** (Eq. A-3)
Card H	Initial Conditions
Card I	Matrix **A** (Eq. A-4)
Card I-1	Matrix **B** (Eq. A-4)
Card J	Measurement Data

Free Field Input Data

The input data is not restricted to any particular format. It consists of a string of values separated by one or more blanks. If all of the remaining values on a single line are zero, they need not be entered. However, zeros within a data string must be entered.

Comment Cards

Comment cards may be inserted anywhere in the data stream after the title card. The comment card must have a C in the first column.

End Card

The last card of the data set must have END beginning in column 1.

The input data for DYNAVAL is listed in this section. The symbol (F) represents a floating point number and (I) represents a fixed point integer.

CARD A Title Card (80 Characters)

CARD B Control Card
Entry 1 I N Total Number of Variables
2 I NG Number of Unknown Forcing Terms
3 I NZ Number of Measurements
4 F Initial Value of the Smoothing Parameter b
5 I ITM Maximum Number of Iterations (GCV)
6 I IOPT
IOPT = 0 Use Specified Initial Conditions
IOPT = 1 Compute Optimal Initial Conditions
7 I KEEP KEEP = 1 Keep the Plot File (.PL1)
8 I IM
IM = 0 Use Exponential Matrix
IM = 1 Read in M and P Directly (Equation A-2).

CARD C Time Increment and Maximum Time
Entry 1 F DELTA Integration Time Increment
2 I NMAX Maximum Number of Timesteps
3 I INC Printout Every INC Timestep
4 I INCP Print to Plot File Every INCP

CARD D Matrix **C** (Diagonal) (End with a –1)
Entry 1 I Location I of Matrix **C**(N × N)
2 I Location I of Matrix **C**
3 F Factor in the (I,J)th Location.
Nonzero Entries only.

CARD E Matrix **K** (End with a –1)
Entry 1 I Location I of Matrix **K**(N × N)
2 I Location J of Matrix **K**
3 F Factor in the (I,J)th Location.
Nonzero Entries only.

CARD F Matrix **P** (End with a –1)
Entry 1 I Location I of Matrix $\mathbf{P}(N \times N_G)$
2 I Location J of Matrix **P**
3 F Factor in the (I,J)th Location.
Nonzero Entries only.

CARD G Matrix **U** (End with a –1)
Entry 1 I Location I of Matrix $\mathbf{U}(N_Z \times N)$
2 I Location J of Matrix **U**
3 F Factor in the (I,J)th Location.
Nonzero Entries only.

CARD H Initial Conditions (End with a –1)
Entry 1 I N Variable Number
2 F x Initial Value

CARD I Data Weighting Matrix **A** (End with a –1)
Entry 1 I Location I of Matrix $\mathbf{A}(N_Z \times N_Z)$
2 I Location J of Matrix **A**
3 F Factor in the (I,J)th Location.
Nonzero Entries only.

CARD I1 Force Weighting Matrix **B*** (End with a –1)
Entry 1 I Location I of Matrix $\mathbf{B}(N_G \times N_G)$
2 I Location J of Matrix **B**
3 F Factor in the (I,J)th Location.
Nonzero Entries only.

CARD J Measurement Data
Entry 1 I Time Step Number
2 F Data
3 F Data
4 F Data
5 F Data
6 F Data
(If more data are required use Card JB)

CARD JB Measurement Data
Entry 1 F Data
2 F Data
3 F Data
4 F Data
5 F Data
(Cards J and JB are required for each time step.)

---

* **NOTE:** This matrix **B** is relative. It will be multiplied by the smoothing parameter b(CARD B).

A.4 EXAMPLE PROBLEM

This problem represents a simple thermal lumped capacitance and resistance model. It will illustrate the basic input to DYNAVAL. The model is shown below.

An unknown heat flux q has been applied to node 3 with an area of 0.4. A temperature measurement has been made at node 1. The differential equations for this model are:

$$0.5T_1 = 1.0(T_2 - T_1)$$

$$1.25T_2 = 1.0(T_1 - T_2) + 0.50(T_3 - T_2) + 4.0(T_4 - T_2)$$

$$0.25T_3 = 0.5(T_2 - T_3) + 0.4q$$

$$0.50T_4 = 4.0(T_2 - T_4)$$

In matrix-vector form, these equations become

$$\begin{bmatrix} 0.50 & 0 & 0 & 0 \\ 0 & 1.25 & 0 & 0 \\ 0 & 0 & 0.25 & 0 \\ 0 & 0 & 0 & 0.50 \end{bmatrix} \begin{Bmatrix} T_1 \\ T_2 \\ T_3 \\ T_4 \end{Bmatrix} = \begin{bmatrix} -1.0 & 1.0 & 0.0 & 0.0 \\ 1.0 & -5.5 & 0.5 & 4.0 \\ 0.0 & 0.5 & -0.5 & 0.0 \\ 0.0 & 4.0 & 0.0 & -4.0 \end{bmatrix} \begin{Bmatrix} T_1 \\ T_2 \\ T_3 \\ T_4 \end{Bmatrix} + \begin{Bmatrix} 0.0 \\ 0.0 \\ 0.4 \\ 0.0 \end{Bmatrix} q$$

The initial temperatures are equal to 10.0. The input data are listed on the following pages. The output from DYNAVAL is also shown.

LISTING OF DYNAVAL INPUT DATA

```
EXAMPLE 1 ** LUMPED CAPACITANCE & THERMAL RESIS-
TANCE MODEL
C ** CARD B — CONTROL CARD
C NODES NG NZ    B     ITERMAX  IOPT  IKEEP
      4  1  1  1.E-5         5     0      1
C ** CARD C - TIME CONTROL
C DELTA NMAX INC INCP
   0.50     20   1      1
```

```
C ** CARD D — MATRIX C DEFINITION
C ILOC JLOC VALUE
   1     1    0.5
   2     2   1.25
   3     3   0.25
   4     4    0.5
  -1
C ** CARD E — MATRIX K DEFINITION
C ILOC JLOC VALUE
   1     1    -1.
   2     2   -5.5
   3     3   -0.5
   4     4   -4.0
   1     2    1.0
   2     1    1.0
   2     3    0.5
   3     2    0.5
   2     4    4.0
   4     2    4.0
  -1
C ** CARD F — PARTICIPATION MATRIX P
C THIS IS FOR FLUX 1 APPLIED TO NODE 3
C NODE FLUX FACTOR
   3     1    0.4
  -1
C ** CARD G — MEASUREMENT MATRIX U
C MEASUREMENT ON NODE 1
C ILOC JLOC FACTOR
   1     1    1.0
  -1
C ** CARD H — INITIAL TEMPERATURES
C  N   X
   1  10.
   2  10.
   3  10.
   4  10.
  -1
C ** CARD I — DATA WEIGHTING MATRIX A
C   I      J    FACTOR
   1      1    1.
  -1
```

```
C ** CARD I1 — FORCE WEIGHTING MATRIX B(RELATIVE)
C   I      J    FACTOR
    1      1    1
   -1
C ** CARD J — TEMPERATURE DATA
C STEP T1
    0       10.
    1     11.09
    2     14.64
    3     19.89
    4     25.38
    5     30.49
    6     34.93
    7     38.62
    8     41.54
    9     43.65
    10    44.97
    11    45.61
    12    45.86
    13    45.95
    14    45.98
    15    45.99
    16    46.00
    17    46.00
    18    46.00
    19    46.00
    20    46.00
END
```

SAMPLE OUTPUT FROM DYNAVAL

CARD A — PROBLEM TITLE
EXAMPLE 1 ** LUMPED CAPACITANCE & THERMAL RESISTANCE MODEL

CARD B - GLOBAL CONTROL
NUMBER OF VARIABLES = 4
NUMBER OF UNKNOWN FORCING TERMS = 1
NUMBER OF MEASUREMENTS = 1
VALUE OF SMOOTHING PARAMETER = 1.000E–05
NUMBER OF ITERATIONS FOR GCV = 5

COMPUTE INITIAL CONDS(IOPT = 0,NO)(IOPT = 1,YES) = 0
KEEP PLOT FILE (= 1,YES) (= 0,NO) = 1
READ IN M AND P DIRECTLY (= 1,YES) (= 0,NO) = 0

CARD C — TIME STEP CONTROL
INTEGRATION TIMESTEP = 0.50000
NUMBER OF TIMESTEPS = 20
PRINTOUT EVERY NPR STEP = 1
PRINT TO PLOTTER FILE EVERY NPLOT STEP = 1

BBB 1.000E–05 TOTAL ERROR 3.051E–04 GCV 1.094E–02
PRESS 2.152E–03
SRSS ERROR 1.747E–02 SRSS FORCE 3.685E+01

BBB 1.031E–04 TOTAL ERROR 4.063E–04 GCV 2.668E–03
PRESS 1.395E–03
SRSS ERROR 2.016E–02 SRSS FORCE 3.682E+01

TIME = 0.0000E+00
VARIABLES
1 1.000E+01 2 1.000E+01 3 1.000E+01 4 1.000E+01
FORCING TERMS
1 8.500E+01

TIME = 5.0000E–01
VARIABLES
1 1.103E+01 2 1.346E+01 3 5.401E+01 4 1.232E+01
FORCING TERMS
1 8.226E+01

TIME = 1.0000E+00
VARIABLES
1 1.468E+01 2 1.934E+01 3 7.209E+01 4 1.782E+01
FORCING TERMS
1 7.322E+01

TIME = 1.5000E+00
VARIABLES
1 1.986E+01 2 2.538E+01 3 7.801E+01 4 2.388E+01
FORCING TERMS
1 6.036E+01
(..........)

```
    TIME = 9.5000E+00
                  VARIABLES
1  4.600E+01  2  4.600E+01  3  4.600E+01  4  4.600E+01
FORCING TERMS
1  -1.199E-03

    TIME = 1.0000E+01
                  VARIABLES
1  4.600E+01  2  4.600E+01  3  4.600E+01  4 +4.600E+01

    THE SQUARE ROOT OF THE SUM OF THE SQUARES
MEASUREMENT  1  SRSS  2.016E-02
BBB  1.031E-04  TOTAL
ERROR  4.063E-04  GCV  2.668E-03
                  PRESS 1.395E-03
    SRSS ERROR 2.016E-02 SRSS FORCE 3.682E+01
```

A.5 FILE DESCRIPTION AND PLOTTING

DYNAVAL creates a post-processing file containing the variables, unknown forces, data, and the measurements. The amount of information saved is controlled by the input variable NPLOT (Card B). The post-processing file will have the same stem as the input file but will have PL1 appended. for example, the input data file, DAVE.DAT will become DAVE.PL1.

The print statements are as follows (FORTRAN):

```
      WRITE(. , 800) TITLE,NODES,NG,NZ
  800 FORMAT(8A10/3I4)
```

The state below are repeated for each timestep:

```
      WRITE(. , 900) TIME
  900 FORMAT(5(1PE14.5))
      WRITE(. , 900) (X(I),I = 1,NODES)
      WRITE(. , 900) (Z(I),I = 1,NZ)
      WRITE(. , 900) (DATA(I),I = 1,NZ)
      WRITE(. , 900) (G(I),I = 1,NG)
```

The data are compared against the variable Z, which may be a combination of the state variables depending on the measurement influence matrix **U**.

DYNAPLOT is a graphics post-plotting program that will plot the time histories of the data contained in the .PL1 file. DYNAPLOT will prompt the user for information. The options include plots of the state variables (X), the forcing function (G), and the data (DATA). Plots comparing the model results (Z) and the data are also possible. The program can also zoom in on user defined sections for a closer inspection of the results.

INDEX

O

P

Q

R

S

NOTICE

This book contains
computer disks.
Remove before charging.

This book contains 1
computer disk(s)
in the pocket inside the
book cover.
Please check the pocket
before discharging.

NOTICE

Warning of Copyright Restrictions

The copyright law of the United States (Title 17, United States Code) governs the reproduction, distribution, adaptation, public performance, and public display of copyrighted material.

Under certain conditions of the law, nonprofit libraries are authorized to lend, lease or rent copies of computer programs to partons on a nonprofit basis and for nonprofit purposes. Any person who makes an unauthorized copy or adaptation of the computer program, or redistributes the loan copy, or publicly performs or displays the computer program, except as permitted by Title 17 of the United States Code, may be liable for copyright infringement.

This institution reserves the right to refuse to fulfill a loan request if, in its judgment, fulfillment of the request would lead to violation of the copyright law.

*****

Lost or Damaged Disks

If the disks have been lost or damaged, inform the service desk at which you charged out the book. In most cases, replacement copies will be made available within 24 hours (on weekends, it may take longer).